Alfred Bennett

The flora of the Alps

Being a description of all the species of flowering plants indigenous to Switzerland.

Vol. 1

Alfred Bennett

The flora of the Alps
Being a description of all the species of flowering plants indigenous to Switzerland. Vol. 1

ISBN/EAN: 9783337197995

Printed in Europe, USA, Canada, Australia, Japan

Cover: Foto ©berggeist007 / pixelio.de

More available books at **www.hansebooks.com**

THE

FLORA OF THE ALPS

VOLUME THE FIRST

THE

FLORA OF THE ALPS

BEING A DESCRIPTION OF ALL THE SPECIES OF FLOWERING

PLANTS INDIGENOUS TO SWITZERLAND; AND OF THE

ALPINE SPECIES OF THE ADJACENT MOUNTAIN

DISTRICTS OF FRANCE, ITALY, & AUSTRIA

INCLUDING THE PYRENEES

BY

ALFRED W. BENNETT

M.A., B.Sc., F.L.S.

LECTURER ON BOTANY AT ST. THOMAS'
HOSPITAL

WITH ONE HUNDRED AND TWENTY

COLOURED PLATES

IN TWO VOLUMES

VOLUME THE FIRST

LONDON

JOHN C. NIMMO

14 KING WILLIAM STREET, STRAND

MDCCCXCVI

Printed by BALLANTYNE, HANSON & CO.
At the Ballantyne Press

INTRODUCTION

THE Flora of the Alps is a subject of never-failing interest even to the casual visitor to the "playground of Europe." Probably nowhere on the face of the globe is to be found, especially in the spring and early summer, a greater wealth of brightly coloured flowers, often growing in enormous masses, festooning the rocks, and making of the alpine pastures a veritable floral carpet; and the interest is greatly increased by the ease with which many of them can be cultivated in our flower-beds or on our rockeries.

The scope of the present work does not fall in exactly with that of any other in the English language. Its object is to provide the tourist with a handbook by which he can recognise the plants which are likely to attract his attention in his Alpine wanderings. There are excellent Floras of Switzerland, of which the best is Gremli's, translated by Paitson (Nutt); but the area of that book is strictly confined to the Republic of Switzerland. There are also Alpine Floras the range of which extends to the adjacent mountain regions, especially Tirol, such as Dalla-Torre's "Tourist's Guide to the Flora of the Alps," translated and edited by the present writer (Sonnenschein); but in this little book "alpine" plants only are enumerated; many lowland species which are altogether unfamiliar to the English tourist being passed over. The range of this latter book does not extend to the Pyrenees; and in neither of those already mentioned are there any illustrations. The illustrated work most familiar to the English botanist is probably Seboth's "Alpine Plants Painted from Nature," also translated by the present writer (Sonnenschein), in which 400 of the most beautiful and interesting plants are admirably delineated in coloured plates. Correvon's pretty little *Flore coloriée de Poche des Montagnes* (Paris, Klincksieck) has a wider range, including also the Pyrenees, and is illustrated by beautifully executed coloured drawings of 180 species; but the total number described is comparatively limited.

In the present work every species of flowering plant which is reported by competent authorities as growing wild in Switzerland is at least named, and a short description given of all except the commonest English plants, which are familiar to every one interested in our wild flowers. But with regard to the other more southern countries named on the title-page, the lowland plants (not found in Switzerland) are not referred to, since they belong to a totally different flora, the Mediterranean; but only those which, from the altitude at which they grow, have a claim to be regarded as alpine.

In choosing the illustrations for the present volumes, an attempt has been made to represent every leading type of alpine flowering plant. At least one plate is given to every natural order which includes any considerable number of alpine species; and in the case of those genera which especially attract the attention of tourists, such as the Pinks, the Saxifrages, the Gentians, and the Primulas, drawings are given of several species. As the

Flora is not limited to Switzerland, a considerable pro-
portion of the illustrations are of natives of other alpine
regions, especially of the Pyrenees.

As the traveller with botanical tastes approaches Switzerland by any of the ordinary routes through Northern
and Eastern France, he notes but little difference in the
vegetation from that with which he is familiar in our
southern counties. A few strangers soon make their
appearance; the spiny heads of *Eryngium campestre*
are seen by the road-sides, and the damp meadows are
coloured by the yellow thistle (*Cirsium oleraceum*). In
the vineyards he would also find some unfamiliar weeds,
such as various species of *Delphinium*. But until he
reaches Mâcon or Pontarlier the general aspects of the
vegetation are the same : it is the flora of North-Western
Europe. In the mountain woods of the French Jura or
of the northern cantons of Switzerland he will come
across much more to attract his attention; such plants
as *Veronica urticæfolia*, *Prenanthes purpurea*, and *Astrantia minor* meet him everywhere; *Prunella grandiflora* is common by the road-sides; species of *Dianthus*
and *Gentiana* become gradually more predominant; while
among the bushes by the road-sides or in damp meadows
he will notice the red-berried elder, *Sambucus racemosa*,
and the feathery *Spiræa Aruncus*. The sylvan flora of
the calcareous Jura chain is especially interesting; here
are found in abundance the alpine honeysuckle, *Lonicera
alpigena*, with its twin red berries, the mezereon, the weird
herb-Paris, *Paris quadrifolia*, the pretty little *Maianthe-
mum bifolium*, the red-berried crowberry, more than one
species of *Phyteuma*, and many another which the inex-
perienced botanist will transfer with delight to his vascu-

lum. But none of these are truly alpine plants ; and it is
not till he ascends much nearer to the eternal snows that
the full glories of the alpine flora burst upon him ;—the
Rhododendrons, the azure of the innumerable Gentians,
the starlike Saxifrages, white and yellow and purple,
the many-hued Anemones, the feathery heads and silvery leaves of *Dryas octopetala*, the *Primulas*, the
Sempervivums, with their webbed leaves and singular
blossoms, and multitudes of others. Even to those who
do not claim to be botanists, it is a day to remember
when the Edelweiss or the auricula is first gathered in
its native haunt, or the tasselled bell of the *Soldanella*
or Snowbell is first seen appearing through the edge of
the snow.

"What is an alpine plant ?" is a question which does
not admit of a very easy answer. It is not uncommon to
find, in sheltered situations, even at very high altitudes,
plants which are but outliers, carrying on a precarious
existence, of species whose home is at a much lower
level ; while, on the other hand, truly alpine species will
often descend, brought down by the glaciers, far below
any line that could be drawn to separate the alpine from
the sub-alpine zone. We regard as an alpine species one
whose proper home, where it thrives best, is near the
eternal snow, or not below the level to which the larger
glaciers descend. The altitude of such a zone will, of
course, vary with the latitude. The adaptability of plants
to endure a climate which is not altogether congenial to
them gives rise to some singular results. We have said
that north of the Alps the flora is essentially that of
North-Western Europe. South of the Alps it is essentially that of the Mediterranean ; this great range of

snow - clad peaks preventing any great migration of southern plants northwards, or of northern plants south-wards, such as has occurred, for example, in North America, where there is no such chain of mountains running east and west between the warmer and colder parts of the Continent. But a mixture of southern and of alpine floras does, nevertheless, take place to a certain extent, especially in the valley of that great south-flow-ing river, the Rhone. Here, as, for example, at Visp or Brieg, you may find growing side by side plants which have ascended the river-basin from the Mediterranean, and plants which have descended the mountain-sides from the Alps.

To the student of geographical botany the absence of certain plants from a given area is not less interesting than the presence of others; and a comparison of the floras of the British Islands and of Switzerland introduces one to many very interesting facts. Of course there is one large element in our flora that is absent from the Swiss—the maritime; with the exception of one or two species, like the thrift, which are found here and there on the summits of the loftiest mountains, and one or two others, as the yellow horned poppy, which finds a con-genial home on the sandy shores of Lake Neuchâtel. There are also a few Arctic species, such as *Primula scotica* and *Saxifraga nivalis*, which occur in our moun-tain districts, but do not get so far south as Central Europe. But the chief element in our flora which would be strange to a Swiss botanist is a Lusitanian one. Owing to the moisture of our climate and the mildness of our winters, many species characteristic of the south-west of Europe extend their range much farther north on

our shores than they are found elsewhere, especially into
Cornwall and Devonshire, South Wales, and the south and
west of Ireland; and this is the case with some of our
most familiar plants. So accustomed are we to the glories
of our gorse-commons and of our hills clad with purple
heather, that we can scarcely believe how limited is the
range of these plants; and it is only gradually that the
English botanist visiting Switzerland begins to realise
that in her woods there are no bluebells; on her banks
no purple foxgloves; that the gorse and the broom are
hardly ever seen; that in her cornfields are no corn-
marigolds; and that from her mountain-sides our bell-
heathers are entirely absent. To the same category
belong also a number of other plants, not so abundant,
but yet familiar enough to the botanist, especially in our
western counties, which he will fail to find in Switzerland, or, if at all, only very rarely. Among these may be
mentioned the Cornish and the Connemara heaths, the
ill-scented *Iris fœtidissima*, the Pennywort, *Cotyledon
Umbilicus*, with its singular peltate leaves, and quite a
number of bog-plants, such as the fragrant bog-myrtle,
the ivy-leaved bellflower, the yellow butterwort, *Pingui-
cula lusitanica*, the delicate bog-pimpernel, and the bog
St. John's wort, *Hypericum elodes*. There are also one
or two arctic aquatic and bog plants abundant with us,
such as *Lobelia Dortmanna* and the bright yellow bog-
asphodel, *Narthecium ossifragum*, which do not reach so
far south as the Alps.

The origin of the alpine flora has been a subject of
much controversy among the authorities in geographical
botany. The natural orders which are especially charac-
teristic of all mountain regions are the Rosaceæ, Ranun-

culaceæ, Saxifragaceæ, Primulaceæ, Campanulaceæ, and
Gentianaceæ; in the flora of the Alps the Compositæ,
Leguminosæ, Cruciferæ, and Caryophylleæ are also
very strongly represented. In a paper read before the
Linnean Society in 1895, but not yet published, the
highest authority on the flora of the Alps, the late Mr.
John Ball, for some years President of the Alpine Club,
gives the geographical distribution of every Swiss plant,
in tables which are the result of an enormous amount of
careful work. He enumerates 2010 species of flowering
plants, in addition to 335 sub-species. Of these, 1117
species, or a little over one-half, belong to the upper zone
of the Alps. The chain on the southern side of the Rhone
Valley is far richer than the Alps of Central and Eastern
Switzerland or of Savoy. In an introduction to this
paper, Mr. Thiselton-Dyer gives an admirable *resumé* of
the views on the origin of the alpine flora advocated
by the two highest authorities, M. Alphonse De Candolle
and Mr. John Ball. Both these experts agree in a recog-
nition of the fact that "there is an element of great anti-
quity in the alpine flora, which cannot be simply accounted
for by a migration from the north during the glacial
epoch." De Candolle points out that some of the most
ancient fragments of the alpine flora, species of *Primula*,
Pedicularis, and *Oxytropis*, are now to be found only on
the southern slopes of the Alps, existing neither in the
interior of Switzerland nor in the North of Europe.
Furthermore, the alpine species of *Campanula* peculiar
to Mont Cenis, the Simplon, and the neighbouring valleys,
are not related to arctic species, but to those of mountain chains to the eastward. Mr. Ball further states
that of the species included in the alpine flora, 17 per

cent. only are common to the Arctic region, while 25
per cent. are common to the Altai range of Northern
Asia.   The subject is, as Mr. Thiselton-Dyer says, a
complex one; but it can hardly be doubted that "the
alpine flora is a very ancient one, and probably a decay-
ing survivor of one of which the extension was at some
former time far more considerable."

The closer resemblance of the flora of the European
Alps to that of the Asiatic Alps than to that of Scan-
dinavia may be partly accounted for by the fact of the
climatic conditions being more nearly alike.   In the
Arctic Zone the plants which blossom in their brief
summer are exposed to almost uninterrupted daylight for
many weeks, during which there are no nocturnal dews.
In the mountain chains of the Temperate Zone, on the
other hand, bright hot days are succeeded by nights of
nearly equal duration, during which the temperature is
often very low and the dews very heavy.   It is obvious,
therefore, that the external characters of alpine and of
arctic plants may differ considerably, and that a large
amount of adaptation may be necessary before an arctic
form could become established in the Alps, while no such
adaptation need accompany the migration of a species
from Central Asia to Central Europe.

Some of the special characteristics of the alpine flora
are familiar to every visitor to the Alps.   Either from
the large size or the bright colour of the flowers, or from
their being grouped together in great masses, they are
far more conspicuous, especially in the spring and early
summer, than is the case in the lowlands.   Although
these brilliant masses of flowers often belong to genera
which are especially alpine, like the Gentians and the

Primulas, yet in other cases it is evidently the result of adaptation to local conditions; whether brought about solely by the operation of natural selection, or partly also by the direct action of external influences, it is not necessary to inquire. A large number of instances could be named in which species belonging to the same genus, and therefore nearly related to one another, differ from one another in having more or less conspicuous flower according as they grow at a higher or a lower elevation. It will be necessary only to remind the botanical reader of such genera as *Anemone, Ranunculus, Dianthus, Prunella, Salvia,* and *Scutellaria.* On the other hand, the number of fragrant plants is probably smaller at high than at low altitudes; the reason being no doubt that, whatever may be the purpose served by the odour of flowers, it would be neutralised by the strong winds which so constantly prevail among the mountains. The explanation usually offered of the bright colour and large size of alpine flowers is that it is necessary for the attraction of the comparatively small number of winged insects which inhabit the regions of the Alps, and which are required for the pollination of the flowers and the fertilisation of the ovules. That the number of insects which could assist in cross-pollination is small in the Alps is undoubted; though Dr. Joly states that numbers are attracted to the glaciers and snowfields in the dusk by the reflected light, and perish there in large numbers. But a doubt is raised as to whether this is the true explanation of the beauty of the alpine flora, from the fact that a comparatively small number of alpine plants are dependent on the ripening of the seeds for their propagation.

But there are other points of structure besides the colour and brightness of the flowers in which alpine species show characteristic differences from their lowland cousins. The summer, the period of the ripening of the seeds, is short; consequently but few species are annual, and therefore entirely dependent on seeds for their propagation. The root-system is, as a rule, strongly developed; both in order to afford sufficient protection against the prevalent strong winds, and for the purpose of drawing as large an amount of nutriment as possible from the scanty soil. On the other hand, in the parts above the surface of the soil the development of woody tissue is often but small. The leaves are protected in various ways against excessive transpiration or evaporation of moisture from the hot summer sun. They are often very crowded (cæspitose), or are covered with a dense felt of hairs; or they are thick and fleshy, their tissue being permeated by canals and chambers containing water. The very early period of flowering of many of the most beautiful species is due to their almost invariably perennial character. The plant has not to spring up from seed and then to produce its buds and flowers; the flowers are already there, formed during the preceding summer, and gradually developing beneath the friendly covering of snow, ready to expand as soon as the April zephyrs call them forth from their hiding-place. Further evidence that the alpine flora is the result of adaptation to climatic conditions is afforded by the fact that, within the bounds of what must unquestionably be regarded as a single species, individuals change in character as we ascend to higher altitudes, the leaves becoming more hairy or more fleshy, and the flowers brighter or larger. Professor G.

Bonnier, who has bestowed on this subject an enormous amount of research, has published, in the *Annales des Sciences Naturelles* for 1894, the results of a very large series of experiments on the cultivation, at different altitudes, of individuals obtained by the division of a single parent stock, which fully confirm these statements. Professor Chodat also, of Geneva, has established in the Jura an alpine botanic garden for the observation of similar phenomena.

It is hoped that, with the present work in his hand, the tourist who is already acquainted with the most familiar English plants will be able to recognise, in the great majority of cases, the plants which he meets with in his alpine wanderings. But he must not expect to be able to do so in all cases. The experienced botanist is often at a loss in distinguishing between two nearly allied species, even with much fuller descriptions before him. The acceptance of the theory of evolution, now all but universal, implies that there may be no hard and fast lines between species, any more than there are between varieties; that a species is not a sharply defined assemblage of individuals which has existed as such from time immemorial; that varieties, species, genera, tribes, orders, may all pass into one another by insensible gradations. There are some genera—among these may be especially mentioned the large families of Brambles, Roses, Hawkweeds, and Willows—in which different authorities differ most widely in their interpretation of what constitutes specific difference; Continental botanists, as a rule, describing a much larger number of species than their English *confrères*. In these cases special manuals must be consulted. And in other instances, as, for example, in the orders Umbelliferæ and Compositæ, and in the genera *Saxifraga, Achillea, Euphrasia,* and *Carex,* a full description of the minute differences between the species would have extended the work to an unwieldy size. The common practice of systematists of giving a *clavis* of the genera in each order and of the species in each genus has been discarded, because this frequently results necessarily in naming only a single character by which species may be distinguished, to the omission of others much more useful to the unpractised botanist. In all the larger genera the species are first of all classified into a number of groups, some conspicuous character being used, such as the colour of the flower, or the arrangement of the flowers or leaves, which does not necessarily imply genetic affinity, and which therefore often separates species which would be placed near to one another in a more scientific flora. We may give as an example of this *Anemone sulphurea*, which is undoubtedly nearly allied to *A. alpina*, the former being found on the granitic, the latter on the calcareous Alps; and yet they are placed in different sections owing to the difference in the colour of their flowers.

With regard to the cultivation of alpine plants in our gardens and rockeries, it must always be borne in mind that their chief enemy in our climate is not cold, but excessive moisture in the winter, from which they are protected in their native country by the friendly covering of snow. As to practical details, we cannot do better than quote the substance of the instructions given by Graf in his Introduction to Seboth's " Alpine Plants " :—

"The great object of the cultivation of alpine plants

should be to reproduce, as nearly as possible, the conditions under which they grow in their native habitat. For
this purpose, it should be observed, in the first place,
whether they are natives of the calcareous or of the
granitic Alps. If they belong to the latter category, it is
quite possible that even a small amount of lime in the
soil would be fatal to their vigorous growth. A suitable
soil having been prepared, they should then be placed in
a position where their roots are not constantly saturated
with moisture. This is best effected by planting them
in a rockery, not too much exposed to the sun, and,
at all events at first, sprinkling them with moisture
several times a day. If planted in pots — which is
generally desirable on first transplantation from their
native mountains — care should also be taken that
the soil, though not suffered to get dry, should be thor-
oughly well drained by broken pottery or other similar
material.

"Many alpine plants will remain for several years in
the same spot, and thrive better when not disturbed.
This is the case with all shrubby species, and with those
whose roots penetrate very deep into the soil, as, for
example, the species of *Astragalus, Oxytropis, Astrantia,*
&c.; others, on the contrary, must be transplanted
every year. In some species, the older portions of
the rhizome die off in consequence of their continued
growth, as may be seen in *Wulfenia carinthiaca,
Armeria alpina,* and many species of *Valeriana* and
Primula."

It has already been remarked that the majority of
alpine plants are not dependent upon the production of
seeds for their natural propagation. Many will, how-

ever, ripen their seeds in our climate. But the greater number will have to be multiplied by division, or, in the case of the more shrubby species, by slips or cuttings.

With a little experience, and probably after a few failures, the tourist will find that he is able, in his own garden, to perpetuate the recollection of many of his most cherished finds. Species of *Saxifraga* and *Sempervivum* are among those which will best repay his attention.

The following are the works which have been chiefly consulted in the preparation of the accompanying Flora :—

GREMLI. "*The Flora of Switzerland.*" Translated by Paitson.

CHRIST. "*Das Pflanzenleben der Schweiz.*"

CORREVON. "*Les Plantes des Alpes.*"

CORREVON. "*Flore coloriée de Poche de la Suisse, &c.*"

DALLA-TORRE. "*Tourist's Guide to the Flora of the Alps.*" Translated by Bennett.

SEBOTH. "*Alpine Plants Painted from Nature.*" Edited by Bennett.

WEBER. "*Die Alpenpflanzen Deutschlands und der Schweiz.*"

NYMAN. "*Conspectus Floræ Europææ.*"

ARDOINO. "*Flore du Département des Alpes Maritimes.*"

GRENIER & GODRON. "*Flore de France.*"

PHILIPPI. "*Flore des Pyrénées.*"

HOOKER. "*The Student's Flora of the British Islands.*"

THURMANN. "*Essai de Phytostatique appliqué à la Chaine du Jura.*"

WOODS. "*The Tourist's Flora.*"

LENTICCHIA. "*Contribuzioni alla Flora della Svizzera italiana.*" (Ticino.) In *Nuovo Giorn. Bot. Ital.* for 1896.

I have also to thank the Council of the Linnean Society for their courtesy in allowing me to see proof-sheets of the unpublished essay by the late Mr. John Ball, to which reference has already been made.

ALFRED W. BENNETT.

6 PARK VILLAGE EAST, REGENT'S PARK,
April 1896.

LIST OF COLOURED PLATES

VOLUME THE FIRST

THE
FLORA OF THE ALPS

Sub-Kingdom I.—ANGIOSPERMS.

Seeds contained in a closed ovary.

Division I.—**DICOTYLEDONS** or **EXOGENS**.

Stem, when woody and perennial, with distinct pith, bark, and annual rings of wood; leaves usually net-veined; sepals, petals, and stamens usually in fours or fives; embryo with two opposite cotyledons.

CLASS I.—THALAMIFLORÆ.

Flowers usually with both calyx and corolla; petals (when present) distinct; stamens springing from beneath the ovary. (Orders I.–XXIV.)

Order I.—RANUNCULACEÆ.

Flowers regular or irregular; stamens indefinite in number; sepals usually 5 or more, deciduous, often coloured; petals usually 5 or more, often minute or 0; carpels usually numerous and distinct, maturing into

A

achenes or follicles. Abundant throughout the temperate and colder parts of the globe.

Tribe CLEMATIDEÆ.—Sepals valvate, coloured. Shrubs with opposite leaves. Genera 1, 2.

1. CLEMATIS, L.

Corolla o; sepals green or white (in European species), usually 4.

The Swiss species are *C. Vitalba*, L., the Traveller's Joy or Old Man's Beard of our hedges, and *C. recta*, L., an erect shrub, with white sepals, hairy at the edge only; growing in thickets. *C. Flammula*, L., a climbing species with white scented flowers, occurs also in the Mediterranean Alps, and in Styria.

2. ATRAGENE, L.

Petals numerous, minute; sepals large, lilac.

A. alpina, L. (Pl. 1), the only species, a beautiful climbing shrub, with large lilac flowers, rarely white, occurs in bushy places in the Alps, Carpathians, and Eastern Pyrenees.

Tribe ANEMONEÆ.—Sepals imbricate; fruit composed of achenes, each with 1 pendulous seed. Herbs. Genera 3–6.

3. THALICTRUM, L.

Sepals imbricate; petals o; fruit a small head of achenes.

The following species are alpine or sub-alpine :— *T. aquilegifolium*, L.; a beautiful plant with stalked carpels, conspicuous violet or lilac stamens, and thrice-ternate leaves; frequent in bushy places; *T. alpinum*, L.; a small high-alpine plant, with the flowers in a simple

terminal raceme, and flower-stalk reflexed after flowering. The remaining alpine or sub-alpine species, often difficult to distinguish, are as follows :—

T. minus, L.; leaves ternate, stem geniculate, flowers in branched panicles. *T. majus*, Jacq.; very much resembling the last, but larger. *T. fœtidum*, L.; the whole plant covered with numerous glandular hairs; local. *T. simplex*, L. (Pyrenees); a dwarf plant, with the flowers almost buried in the leaves. *T. saxatile*, DC. (Valais), (*T. alpestre*, Gaud.); a small species with leaves crowded towards the middle of the stem. *T. alpicolum*, Jord. (*Bauhini*, Crntz.); (very local), with numerous flowers in dense panicles. *T. augustifolium*, L. (Jura, Southern Tirol, Dauphiny); flowers in a pyramidal panicle, stamens pendant.

T. flavum, L., our English Meadow Rue, is common in wet places; and the nearly allied *T. exaltatum*, Gaud., is found on the shores of Lake Lugano. *T. macrocarpum*, Gren., with swollen carpels, and *T. tuberosum*, L., with yellowish flowers and large persistent sepals, occur also in the Pyrenees.

4. ANEMONE, L.

Sepals 4–20, imbricate, petaloid; petals 0; leaves all radical; stem with three or more leaf-like bracts.

Besides our own *A. nemorosa*, L., Wood Anemone or Windflower, and *A. Pulsatilla*, L., Pasque-flower, with purple flowers, neither of which is alpine, the following species occur in Switzerland :—

A. Flowers in umbels of 2–10, surrounded by an involucre :—*A. narcissiflora*, L.; flowers 3–6, white; Alps, Jura, Pyrenees.

B. Flowers solitary; sepals white, or pink on the outside; involucral bracts usually resembling the leaves:— *A. baldensis*, L.; style of fruit short, not feathery, sepals hairy beneath, usually 9; Alps, Pyrenees, Dauphiny. *A. sylvestris*, L.; calyx hairy, leaves 5-partite, carpels very woolly, style not feathery; Dauphiny, Pyrenees. *A. trifolia*, L.; sepals glabrous beneath, usually 6, style not feathery; Tirol, Carinthia. *A. alpina*, L. (Pl. 2); style of fruit long, feathery, sepals 6–9, lanceolate; common. *A. Burseriana*, Scop.; resembling the last, but sepals broadly ovate, usually 7; frequent.

C. Sepals purple or violet:—*A. Halleri*, All.; flower solitary, lilac, whole plant covered with long silky hairs; local; Nicolaithal, Dauphiny. *A. vernalis*, L.; flowers light violet or nearly white, solitary, involucral leaves sessile, united into a sheath. *A. montana*, Hoppe; flowers solitary, dark violet, drooping, leaves glabrous, with linear lobes; local. *A. hortensis*, L.; the origin of our garden Anemone; flowers variable in colour, erect, sepals glabrous outside, carpels woolly; Pyrenees.

D. Flowers yellow:—*A. sulphurea*, L.; flowers solitary, sulphur-yellow, style of fruit long, feathery; high Alps. *A. ranunculoides*, L.; flowers 1–3, small, yellow, sepals 5–8, pubescent, plant glabrous; Jura, Dauphiny, Vosges, Pyrenees.

E. Flowers small, blue, leaves 3-lobed:—*A. Hepatica*, L., the Hepatica; Alps, Jura, Pyrenees.

5. ADONIS, L.

Petals 5–16, conspicuous; leaves divided into very narrow segments. Not alpine.

The Pheasant's-eyes are not alpine plants. *A. autum-nalis*, L., with purple-red petals (often with a black spot); *A. æstivalis*, L., with light red (rarely yellow) petals and glabrous sepals; *A. flammea*, Jacq., with light red petals and hairy sepals; *A. vernalis*, L., with pale yellow petals and pubescent sepals; and *A. pyrenaica*, DC., with yellow petals and glabrous sepals, occur as weeds in cultivated land in the Pyrenees and extreme south of Switzerland.

6. MYOSURUS, L.

Petals small, tubular; sepals 5, spurred; fruit a long spike of densely packed achenes. Not alpine.

M. minimus, L., Mouse-tail, the British and only European species, an inconspicuous plant in cornfields and sandy places.

Tribe RANUNCULEÆ.—Sepals imbricate; fruit composed of achenes, each with one erect seed. Genus 7.

7. RANUNCULUS, L.

Petals usually 5, white or yellow.

A. Flowers yellow; petals 8–12:—*R. Ficaria*, L.; our common Pilewort or Lesser Celandine; leaves nearly round, cordate; everywhere.

All our common meadow or cornfield species of Buttercup or Crowfoot grow also in Switzerland, viz. :—*R. acris*, L.; *repens*, L.; *bulbosus*, L.; *hirsutus*, Curt.; *arvensis*, L.; and *auricomus*, L.; also the marsh species, *R. sceleratus*, L.; the very handsome Great Spearwort, *R. Lingua*, L.; and the Lesser Spearwort, *R. Flammula*, L., both with narrow, undivided leaves. The following are alpine or sub-alpine:—

B. Flowers yellow; petals 5; leaves narrow, undivided: —*R. gramineus*, L.; flowers 1-7, all the leaves linear-lanceolate; Southern Switzerland, Pyrenees, rare. *R. ophioglossifolius*, Willd.; flowers numerous, lower leaves cordate, upper elliptical; Pyrenees.

C. Flowers yellow; petals 5; leaves peltate or reniform:—*R. Phthora*, Crntz.; leaves reniform, radical leaves on long stalks, root-fibres tuberous; local. *R. Thora*, L.; leaves reniform, radical leaves 0, root-fibres tuberous; Alps, Jura, Pyrences, local. *R. Schottii*, Dalla Torre, Styria, scarcely differs.

D. Flowers yellow; petals 5; all the leaves more or less divided:—*R. lanuginosus*, L.; leaves palmately 3-5-partite, the whole plant covered with yellowish hairs; mountain forests; Switzerland, Jura, Pyrenees, rare. *R. pygmæus*, Whlb.; stem with one leaf and one very small flower ($\frac{1}{4}$ in. diam.); high; Tirol, Salzburg, Carinthia, rare. *R. nemorosus*, DC. (*aureus*, Schleich.; *sylvaticus*, Gren.); flowers large ($\frac{1}{2}$ to 1 in.), golden yellow, fruit-stalk furrowed, beak of fruit strongly recurved; woods. *R. polyanthemos*, L.; resembling the last, but flower smaller, beak shorter; woods, rare; Chur. *R. Villarsii*, DC. (*aduncus*, G. and G.); stem 6-18 in., beak of fruit hooked; Southern Switzerland, Carinthia, Pyrenees. *R. Gouani*, Willd.; stem 4-15 in., stem-leaves amplexicaul, beak of fruit long, curved; Pyrenees. *R. Breynianus*, Crntz.; stem $1\frac{1}{2}$-6 in., radical leaves with sharply toothed segments; Switzerland. *R. minutus*, Leyb.; stem 1-flowered, sepals narrow, as long as the petals, leaves $\frac{1}{4}$ in. diam.; Carinthia, rare. *R. mon-*

III.—RANUNCULUS PARNASSIFOLIUS.

tanus, Willd.; stem 1-flowered, 2-6 in., sepals pubescent, leaves hairy; pastures; Alps, Jura, Pyrenees. *R. carinthiacus*, Hoppe; stem 1-flowered, delicate; sepals shorter than petals; Eastern Alps.

E. Flowers pink; petals 5:—*R. roscus*, Heg.; calyx covered with rough hairs; glaciers; Switzerland, rare.

F. Petals more than 5, white with yellow base:—*R. rutæfolius*, L. (*Callianthemum rutæfolium*, Mey.); petals 6-12, obovate, radical leaves doubly pinnate; high; Switzerland, Dauphiny, Pyrenees. *R. anemonoides*, Zahl.; petals 9-20, linear-cuneate, radical leaves biternate; high; Southern Tirol, Styria, rare.

G. Petals 5, white or slightly pink; leaves entire:—*R. pyrenæus*, L.; radical leaves linear-lanceolate, stem-leaves not amplexicaul; Alps, Pyrenees. *R. parnassifolius*, L. (Pl. 3); radical leaves cordate-ovate, stem-leaves amplexicaul; high; Southern Switzerland, Dauphiny, Pyrenees, rare. *R. amplexicaulis*, L.; radical leaves ovate-lanceolate, stem-leaves amplexicaul; Pyrenees.

H. Petals 5, white or slightly pink; leaves more or less divided:—*R. glacialis*, L.; calyx woolly with rough hairs, radical leaves ternate; very high; Switzerland, Dauphiny, Pyrenees. *R. aconitifolius*, L.; stem 1-3 ft., flexuous; flowers $\frac{1}{2}$-$\frac{3}{4}$ in.; petals with a nectariferous scale at the base, leaves divided to the base; high, damp; Alps, Jura, Vosges, Pyrenees. *R. Pacheri*, Dalla Torre; flowers smaller, plant more hairy; Carinthia. *R. platanifolius*, L.; smaller than *aconitifolius*, lower leaves divided $\frac{2}{3}$, upper nearly entire; Alps, Jura, Pyrenees. *R. Seguieri*, Vill.; plant 2-4 in., villous, 1-3-leaved, leaves deeply palmate; Dauphiny, Jura, Carinthia, local. *R. alpestris*, L.; stem 2-4 in., radical leaves 3-5-cleft, beak of fruit hooked; Alps, Jura, Dauphiny, Pyrenees, Carpathians. *R. Traunfellneri*, Hoppe; resembling the last, but beak of fruit only curved, leaves more deeply cleft; Tirol, Carniola, rare. *R. crenatus*, W.K.; radical leaves entire, serrate, petals roundish-ovate, beak of fruit curved; Styria. *R. bilobus*, Bert.; radical leaves entire, crenate, petals cuneate-ovate, beak hooked; Southern Tirol, rare.

I. Petals 5, white; aquatic plants (*Batrachium*):—Most of the British species of Water Crowfoot, viz., *R. aquatilis*, L., with its numerous sub-species; *fluitans*, Lamk.; *circinatus*, Sm., are found also in Switzerland.

Tribe HELLEBOREÆ.—Sepals imbricate; fruit composed of a few many-seeded follicles.   Genera 8–17.

8. CALTHA, L.

Flowers regular; sepals petaloid; petals 0.   Not alpine.

Our English Marsh Marigold, *C. palustris*, L., is common throughout Switzerland and the Pyrenees.

9. TROLLIUS, L.

Flowers regular; sepals 5–15, yellow; petals small, entire.

The beautiful Globe-flower, *T. europæus*, L., a native of our northern counties, is abundant in sub-alpine pastures.

10. ERANTHIS, L.

Flowers regular; sepals petaloid, deciduous; petals small, 2-lipped.

The Winter Aconite, *E. hyemalis*, Salisb., with yellow sepals and petals, occurs rarely in bushy places in Jura and Vosges.

11. HELLEBORUS, L.

Flowers regular; sepals green or white, persistent; petals small, 2-lipped.

H. viridis, L., Bear's-foot, with annual leafy stems and green sepals, is found in pastures in Switzerland and Pyrenees. *H. fœtidus*, L., Stinking Hellebore, with perennial leafy stems and green sepals, is very abundant in stony pastures in Western Switzerland (Jura) and Pyrenees. *H. niger*, L., Christmas Rose, with white sepals and leafless stem, occurs locally on calcareous soil in Switzerland, especially in Ticino.

12. NIGELLA, L.

Flowers regular; sepals blue; petals 2-lipped, nectariferous; follicle terminated by a long curved style; leaves divided into linear segments. Not alpine.

Three species of Love-in-a-puzzle, *N. arvensis*, L., *sativa*, L., and *damascena*, L., are weeds in cultivated land in Pyrenees; and *N. arvensis*, distinguished by the absence of an involucre, also in Southern Switzerland.

13. ISOPYRUM, L.

Sepals 5, petaloid, deciduous; petals 5, minute, cornucopia-shaped; fruit of 1–3 follicles.

I. thalictroides, L., with panicles of small white flowers and delicate twice-ternate leaves; in mountain copses in Savoy, Dauphiny, and Pyrenees.

14. AQUILEGIA, L.

Flowers regular; sepals coloured; petals large, spurred; follicles 5.

A. Spur of the petals rolled up at the apex:—Our English Columbine, *A. vulgaris*, L., is common throughout Switzerland and Pyrenees. A variety, *A. atrata*, Koch, with dark violet flowers, is found in bushy places in the Alps. *A. Hænkeana*, Koch, with very large violet flowers, is a native of Carinthia, Carniola, and Tirol.

B. Spur of petals straight or slightly curved:—*A. alpina*, L., resembling *A. vulgaris*, but with larger flowers, is rare in bushy places in Switzerland, Dauphiny, and Pyrenees. *A. pyrenaica*, DC. (Pl. 4), with a slender spur and smaller leaves, is a native of Pyrenees and Southern Switzerland. *A. Einseleana*, F. Schultz, with viscid stem and hairy spur; and *A. thalictrifolia*, Sch. and K., with glabrous spur and linear-lanceolate leaf-segments, are rare plants in Tirol, and the former also in Carinthia.

15. DELPHINIUM, L.

Flowers irregular; sepals and petals coloured and spurred.

D. elatum, L., with azure-blue sepals, rough grey petals with a long narrow spur, and 3–4 carpels, is found locally in dry sub-alpine meadows in Switzerland and Pyrenees; and *D. tirolense*, Kern., with dark blue sepals and petals and narrower leaf-segments, in Northern Tirol. Our English Larkspur, *D. Consolida*, L., with one carpel, is a cornfield weed in Switzerland; and several other species in Pyrenees.

16. ACONITUM, L.

Flowers irregular; sepals and petals coloured; dorsal sepal large, hooded; petals small, clawed.

IV.—AQUILEGIA PYRENAICA.

Several species of Monkshood are common sub-alpine plants.

A. Sepals blue, violet, or white; spur hooked or slightly curved:—*A. Napellus,* L., a glabrous plant, with bright blue (occasionally white) flowers, is abundant in alpine pastures in Switzerland and Pyrenees. *A. paniculatum,* Lam., with a looser raceme, hairy stem, and dark bluish-violet flowers, occurs in damp bushy places in Switzerland, Dauphiny, and Pyrenees; and *A. variegatum,* L., with spotted sepals and erect petals, in Switzerland. *A. Störkeanum,* Rchb., is probably a hybrid.

B. Sepals yellow or yellowish-white; spur coiled:—*A. Lycoctonum,* with pale yellow flowers, palmate leaves, and glabrous follicles, is common at low elevations; and *A. Anthora,* L., with more erect darker flowers and hairy follicles, at higher elevations, in the Alps and Pyrenees. *A. ranunculifolium,* L., with palmate leaves and hood nearly three times as high as broad; and *A. commutatum,* Dall. Torr., resembling the last, but with shorter leaflets, inhabit the calcareous Alps. *A. pyrenaicum,* L., with velvety golden-yellow flowers and very fleshy leaves, is found in Pyrenees.

17. PÆONIA, L.

Flowers solitary, regular; sepals 5, green, persistent; petals numerous, very large.

Three species of Peony are occasionally met with in bushy places in Styria, Tirol, Carinthia, and Southern Switzerland, viz.:—*P. corallina,* Retz., with 4–5 carpels horizontal in fruit; *peregrina,* Mill., with 2–3 erect carpels

(Monte Generoso); and *pubens*, Sm., resembling the last, but with narrower leaflets.

Tribe ACTÆEÆ.—Flowers nearly regular; fruit a berry.

18. ACTÆA, L.

Flowers small; sepals 3–5, petaloid; petals small or o.

A. spicata, L., Bane-berry, with black many-seeded berries, small white flowers in a simple raceme, and ternately compound leaves, is met with in copses in Switzerland and Pyrenees.

Order II.—BERBERIDEÆ.

Flowers regular, trimerous; stamens definite, opposite the petals; anthers opening by recurved valves; carpel 1; fruit a 1–2-seeded berry or drupe. A small order of shrubs or small trees, belonging mostly to the temperate regions.

1. BERBERIS, L.

Sepals 8–9, petaloid; petals 6; stamens 6; fruit a 1–2-seeded berry. Not alpine.

B. vulgaris, L., our Barberry, grows in thickets throughout the district, and is especially abundant in Eastern Switzerland (Grisons).

2. EPIMEDIUM, L.

Sepals, petals, and stamens 4 each; flowers in a very loose raceme.

E. alpinum, L.; flowers reddish-brown, leaves biternate, leaflets cordate; Tirol, Carniola; not wild elsewhere.

Order III.—NYMPHÆACEÆ.

Flowers regular; stamens numerous; stigma sessile on the many-celled ovary. A small order of aquatic plants, usually with large handsome flowers and floating leaves.

1. NYMPHÆA, L.

Sepals 4; petals numerous, gradually passing into the stamens.

N. alba, L., the White Waterlily; in lakes and ponds.

2. NUPHAR, Sm.

Sepals 5–6, yellow; petals and stamens numerous.

N. luteum, Sm., our Yellow Waterlily; in lakes and ponds. *N. pumilum*, Sm., with much smaller flowers, unguiculate petals, and oblong leaves; rare; in mountain lakes in Switzerland, Vosges, Tirol, and Carinthia.

Order IV.—PAPAVERACEÆ.

Flowers regular; sepals 2, deciduous; petals 4; stamens indefinite. Herbs with a milky or coloured juice (latex). A small order, belonging chiefly to the temperate zone.

1. PAPAVER, L.

Ovary 1-celled, with internal projections to which the very numerous ovules are attached, opening by small pores beneath the sessile persistent stigma.

Our three species of Poppy, *P. Rhœas*, L., *dubium*, L. (including *Lecoquii*, Lam., and *collinum*, Bog.), and *Argemone*, L., are cornfield weeds; also *P. hybridum*, L., with a short ovoid capsule covered with stiff very spreading hairs, in Southern Switzerland and Pyrenees. *P. alpinum*, L., the Alpine Poppy (Pl. 5), one of the most beautiful plants of the Alps, with white or less often yellow flowers, is not uncommon on calcareous or slaty mountain-sides in Switzerland, Dauphiny, and Pyrenees (*P. pyrenaicum*, Willd.; *rhæticum*, Ler.).

2. MECONOPSIS, Vig.

Style distinct; stigmas 4; ovary not septated; ovules parietal.

M. cambrica, Vig., the Welsh Poppy, with pale yellow flowers; mountain woods in Pyrenees.

3. CHELIDONIUM, L.

Ovary 1-celled; capsule elongated. Not alpine.

C. majus, L., Celandine; hedges and waste places; common.

4. GLAUCIUM, Scop.

Ovary 2-celled; capsule greatly elongated; leaves thick, fleshy. Not alpine.

G. luteum, Scop. (*flavum*, Crntz.), the Yellow Horned Poppy; shores of Lake Neuchatel; *G. corniculatum*, Curt. (*phœniceum*, Crntz.); petals scarlet, with a black spot at the base; fields; Southern Switzerland, Pyrenees.

5. HYPECOUM, L.

Stamens 4; styles 2; capsule greatly elongated, jointed.

H. procumbens, L.; with yellow flowers; Pyrenees.

Order V.—FUMARIACEÆ.

Flowers irregular; sepals 2, deciduous; petals 4, one or two of them gibbous or spurred; stamens 6, in two bundles of 3 each; ovary 1-celled. A small order; chiefly of western distribution. Not alpine.

1. FUMARIA, L.

Flowers small, usually pink; one only of the outer petals spurred; fruit globose, 1-seeded.

Our English species of Fumitory; *F. officinalis*, L., *parviflora*, Lam., and *Vaillantii*, Lois, are weeds in cultivated and waste land; and *F. capreolata*, L., locally in the neighbourhood of Geneva, Lausanne, and Eastern Pyrenees.

2. CORYDALIS, DC.

Flowers usually larger, and white, yellow, or violet; one petal only spurred; fruit a 2-valved capsule, many-seeded.

A. Root tuberous; flowers violet, purple, or white:— *C. cava*, Schw., with thick curved spur; *C. fabacea*, Pers., with straight spur and nearly sessile flowers; and *C. solida*, Sm., with straight spur and flowers on longer stalks; in hilly woods in Jura, Southern Switzerland, Vosges, Dauphiny, and Pyrenees.

B. Root fibrous:—*C. lutea*, DC., with large yellow flowers; Ticino, very local. *C. claviculata*, DC.; a very pretty climbing plant, with small nearly white or very pale yellow flowers; Pyrenees. *C. ochroleuca*, Koch, flowers yellowish-white with dark tips, is reported from Monte Generoso.

Order VI.—CRUCIFERÆ.

Flowers regular, mostly white or yellow, in bractless racemes; sepals 4; petals 4; stamens 6, of which 2 have shorter filaments than the other 4 (rarely only 4); fruit a siliqua or silicule. A very large order of herbaceous plants, most abundant in the Temperate and Arctic regions of Europe and Asia.

Tribe ARABIDEÆ.—Seed-vessel elongated, much longer than broad (a siliqua); seeds in one or two rows; radicle accumbent. Genera 1–7.

1. MATTHIOLA, Br.

Stigma erect or decurrent on the style; flowers large, lilac.

M. valesiaca, Boiss., a woolly plant with flowers a dirty violet, is recorded from Southern Switzerland and Tirol; but no species of Stock are alpine plants.

2. CHEIRANTHUS, L.

Stigma terminal; flowers large, yellow or variegated; lateral sepals saccate at the base. Not alpine.

C. Cheiri, L., Wall-flower; on walls and rocks in Southern Switzerland, often introduced.

3. NASTURTIUM, Br.

Stigma terminal; flowers yellow or white; sepals equal at the base; seed-vessel turgid; seeds in two rows.

N. pyrenaicum, R.Br., the only sub-alpine species,

with yellow flowers, and auricled pinnatifid leaves with linear segments, grows on dry mountain-sides in Switzerland. The other Swiss species are also English, and are aquatic or moisture-loving plants of the lowlands, viz.: *N. officinale*, Br., Watercress, with white flowers; and *N. sylvestre*, Br.; *palustre*, DC.; and *amphibium*, Br.; all with small yellow flowers.

4. BARBAREA, Br.

Stigma terminal; flowers yellow; sepals equal; petals clawed; seed-vessel 4-angled; seeds in one row. Not alpine.

B. vulgaris, R.Br., Winter Cress, with its sub-species, is a very common hedgerow plant.

5. ARABIS, L.

Stigma terminal; flowers usually white; seed-vessel linear, compressed; seeds in one row.

A. Flowers blue:—*A. cærulea*, Hænke; a very pretty plant of the High Alps and Dauphiny, with pale blue flowers and long erect seed-vessels.

B. Flowers pale yellow; stem-leaves amplexicaul, auricled:—*A. perfoliata*, Link (*Turritis glabra*, L.); plant nearly glabrous, stem-leaves sagittate, seeds not winged; stony places. *A. Turrita*, L.; plant pubescent, seeds winged; stony places.

C. Flowers cream-coloured:—*A. stricta*, Huds.; leaves thick, shining, stem-leaves not amplexicaul; Salève, Dauphiny, Pyrenees.

D. Flowers pink or lilac:—*A. arenosa*, Scop.; calyx

gibbous, radical leaves pinnatifid; damp sandy places. *A. ovirensis*, Wulf.; radical leaves simple, on long stalks; Carinthia.

E. Flowers white; stem-leaves cordate or sagittate, amplexicaul:—*A. alpina*, L.; stem with barren shoots, two of the sepals saccate at the base; rocky places, common. *A. saxatilis*, All.; stem without barren shoots, petals ⅛–¼ in. long; local. *A. brassicæformis*, Wallr.; leaves thick, fleshy, stem-leaves amplexicaul; stony places; Alps, Vosges, Jura, Dauphiny, Pyrenees, rare. *A. auriculata*, Lam.; stem-leaves amplexicaul, raceme flexuose; Jura, Dauphiny, Pyrenees, rare. *A. hirsuta*, Scop.; stem and leaves hispid; and *A. sagittata*, DC., stem-leaves cordate-sagittate with pointed auricles; roadsides in the lowlands.

F. Flowers white; stem-leaves narrowed or rounded at the base, not amplexicaul:—*A. serpyllifolia*, Vill.; stem flexuose, all the leaves entire; Switzerland, Jura, Dauphiny, Pyrenees. *A. bellidifolia*, Jacq.; radical leaves dentate, stem-leaves entire, semi-amplexicaul; moist; Alps, Pyrenees, Carpathians. *A. pumila*, Jacq.; radical leaves dentate, stem-leaves entire, sessile, not amplexicaul; Alps, Dauphiny, Carpathians. *A. petræa*, Lam.; very dwarf, radical leaves dentate, stem-leaves entire, stalked; Styria. *A. Halleri*, L.; radical leaves lyrate-pinnatifid, stem-leaves dentate; Engadine, Simplon, rare. *A. vochinensis*, Spreng.; stem stoloniferous; Tirol, Carinthia. *A. alpestris*, Schleich. (*ciliata*, Koch); stem straight, fruit-stalk erect, leaves dentate, hairy or ciliate; Alps, Jura, Dauphiny, Pyrenees. *A. muralis*, Bert.; stem-leaves rounded at the base, plant very hairy; walls and rocks in the lowlands.

VI.—CARDAMINE TRIFOLIA.

6. CARDAMINE, L.

Stigma terminal; flowers pink or white; seed-vessel flat, with elastic valves; sepals equal at the base.

A. Flowers large, usually lilac; not alpine:— *C. pratensis*, L., Cuckoo-flower, Lady's Smock; very common in damp meadows.

B. Flowers white; not alpine:—*C. hirsuta*, L., and *sylvatica*, L., with very small flowers and pinnate radical leaves, the former usually with only 4 stamens; in shady places. *C. amara*, L., Bitter Cress; with large creamy-white flowers and pinnate leaves; river-sides, scarce; and *C. impatiens*, L.; leaves with stipule-like fringed auricles; shady places, local.

C. Flowers white; alpine plants :—*C. alpina*, Willd.; stem 1–2 in. high, leaves ovate, entire; frequent. *C. trifolia*, L. (Pl. 6); radical leaves tripartite, on long stalks, stem-leaves 0; Jura, Tirol, Styria, Carinthia. *C. Plumierii*, Vill.; lower leaves reniform, cordate, upper leaves with 3 or 5 segments; Dauphiny. *C. resedifolia*, L.; radical leaves oblong, entire, on long stalks, stem-leaves pinnatifid; moist rocks; Alps and Pyrenees, rare. *C. asarifolia*, L.; stem 6–12 in., flowers in a corymbose panicle, leaves reniform, toothed; Southern Switzerland, rare.

7. DENTARIA, L.

Stigma terminal; flowers large, white, purple, or yellow; seed-vessel flat, with elastic valves; funicle of seed dilated; stem-leaves often opposite or in whorls.

All the species of Toothwort are sub-alpine plants, growing in moist or shady places :—*D. enneaphyllos*, L.;

flowers yellowish-white, stem-leaves in whorls of three; Tirol. *D. digitata*, Lam.; flowers pink or lilac, leaves alternate, lower with 3, upper with 5 segments; Switzerland, Tirol, Pyrenees. *D. alternifolia*, Hausm.; flowers yellowish-white, leaves alternate, with 3 segments; Tirol, Styria. *D. intermedia*, Sond.; flowers lilac or white, leaves with 5 segments; Southern Tirol, rare. *D. bulbifera*, L., Coralwort; flowers lilac or white, lower leaves pinnatifid, upper undivided, with axillary bulbs; Switzerland, Dauphiny. *D. polyphylla*, W. & K.; flowers yellowish-white, 7-14, all the leaves pinnatifid; rare; Switzerland, Carinthia. *D. pinnata*, Lam.; flowers white or lilac, 3-7, all the leaves pinnatifid; Switzerland, Pyrenees.

Tribe SISYMBRIEÆ.—Seed-vessel elongated, seeds usually in one row; radicle incumbent, straight, planoconvex; flowers white, yellow, or lilac. Genera 8-12.

8. SISYMBRIUM, L.

Plant glabrous or with simple spreading hairs; leaves usually toothed or pinnatifid; flowers in loose racemes, usually yellow. Not alpine.

The following British species occur in corresponding situations in Switzerland:—*S. Thaliana*, Koch, with very small white, and *S. Alliaria*, Scop., Jack-by-the-hedge, Sauce-alone, with larger white flowers; *S. officinale*, Scop., Hedge Mustard, with very small yellow, and *S. Irio*, L., and *S. Sophia*, L., with larger yellow flowers. Also the following in Southern Switzerland:—Flowers yellow:—*S. austriacum*, Jacq., leaves runcinate;

S. strictissimum, L., leaves entire or glandular-dentate; *S. Columnæ*, Jacq. (*orientale*, L.), river-sides, Lugano; and *S. altissimum*, L., with pale yellow flowers and very spreading sepals. Also *S. supinum*, L. (*Braya supina*, Koch), with very small white flowers solitary in the axils of the leaves; and *S. pinnatifidum*, DC. (*Braya pinnatifida*, Koch), with spotted sepals; in rocky places; Pyrenees, Dauphiny, and Southern Switzerland. *S. Sinapistrum*, Crntz. (*pannonicum*, Jacq.), closely resembles *officinale*, with longer pods.

9. BRAYA, Sternb.

Seed-vessel nodose, cylindrical; leaves undivided; flowers small, white; seeds in two rows.

B. alpina, Sternb.; stem 1–3 in. high; raceme very short and crowded; Alps, Carinthia, rare. *B. pinnata*, Sternb.; leaves linear-lanceolate, entire or with a very few teeth, flowers crowded, corymbose, violet when dry; very rare; Tirol, Carinthia (Gross-Glockner).

10. HUGUENINIA, Rchb.

Seed-vessel cylindrical, 2-edged; leaves pinnatifid; flowers yellow.

H. tanacetifolia, Rchb. (*Sisymbrium tanacetifolium*, L.); segments of leaves lanceolate, toothed; rare; St. Bernard, Arollathal, Dauphiny, Pyrenees.

11. ERYSIMUM, L.

Plant hoary with adpressed forked hairs; sepals erect; leaves narrow, entire; flowers yellow or white, often fragrant.

The species of Treacle Mustard are mostly lowland plants, growing on walls, waste ground, &c. Several species with yellow flowers occur in Southern and Western Switzerland, Pyrences, &c., and are difficult to distinguish from one another, viz.: *E. cheiranthoides*, L., with small flowers on long stalks; *E. virgatum*, Roth., with larger flowers and leaves almost entire; *E. strictum*, Wett., with darker flowers and sinuate-dentate leaves; *E. ochroleucum*, DC., with linear-lanceolate slightly toothed leaves; Jura; also *E. perfoliatum*, Crntz. (*Coringia orientalis*, Rchb.), with white flowers and amplexicaul stem-leaves. *E. pumilum*, Gaud. (*helveticum*, Koch), is an alpine plant with small fragrant yellow flowers, and sepals very saccate at the base, found at high elevations in the Alps and Pyrences.

12. Hesperis, L.

Plant pubescent with spreading hairs; leaves entire; sepals saccate at the base; flowers large. Not alpine.

H. matronalis, L., Dame's Violet, with white or lilac flowers, fragrant in the evening; in hedges, woods, &c., throughout the South.

Tribe Brassiceæ.—Seed-vessel elongated; seeds in 1–2 rows; radicle incumbent, longitudinally folded or concave. Genera 13–16.

13. Brassica, L.

Seeds in one row; sepals erect; flowers yellow. Not alpine.

B. campestris, L., the Wild Turnip, is the only undoubtedly wild species in Switzerland.

14. SINAPIS, L.

Seeds in one row; sepals spreading; flowers yellow. Not alpine.

S. nigra, L., Black Mustard; *S. alba*, L., White Mustard; *S. arvensis*, L., Charlock; and *S. Cheiranthus*, M.K., with very deeply divided leaves; are all common weeds in cultivated ground.

15. DIPLOTAXIS, DC.

Seeds in two rows; sepals spreading; flowers yellow; leaves pinnatifid. Not alpine.

D. muralis, DC., with hispid leafy stem; *D. tenuifolia*, L., with much larger flowers; in waste places.

16. ERUCASTRUM, Pr.

Seeds in one row; otherwise like *Diplotaxis*. Not alpine.

E. obtusangulum, Rchb., with amplexicaul leaves; *E. Pollichii*, Schp., with smaller pale yellow flowers and leaves not amplexicaul; and *E. incanum*, Koch (*Hirschfeldia adpressa*, Mœnch.), with small yellow flowers and short cylindrical seed-vessel, the upper leaves lanceolate, undivided; are Southern lowland plants.

17. ERUCA, DC.

Seeds in two rows; sepals equal at the base; seed-vessel cylindrical; stigma bilobed. Not alpine.

E. sativa, Lam.; flowers dirty-white veined with violet, leaves pinnatifid; slopes; Rhone valley.

Tribe ALYSSINEÆ.—Seed-vessel short (silicule), not compressed or jointed; seeds in two rows; radicle accumbent; flowers white or yellow. Genera 18–25.

18. ALYSSUM, L.

Stem leafy; leaves entire; flowers small, white or yellow; sepals equal, entire or bifid; filaments toothed or winged; seed-vessel cylindrical, few-seeded.

A. alpestre, L.; with small pale yellow petals, scarcely longer than the sepals, and woody stem; Nicolaithal, Pyrenees, Dauphiny, rare. *A. Wulfenianum*, Bernh.; with golden-yellow flowers and sharp teeth on the filaments; Switzerland, Tirol, Styria, Carinthia, rare. *A. Rochelii*, Andrz.; with golden-yellow flowers and blunt filament-teeth; Carinthia, also in rocky places in the Pyrenees. *A. montanum*, L., a shrubby plant with larger yellow flowers; and *A. perusianum*, Gay, *A. spinosum*, L., and *A. pyrenaicum*, L., with white flowers, the last two spiny; Pyrenees. *A. calycinum*, L., flowers yellow, is a weed in cultivated land. *A. incanum*, L. (*Berteroa incana*, DC.); flowers white, plant grey with stellate hairs; road-sides; Southern Switzerland, rare.

19. COCHLEARIA, L.

Sepals short, equal, spreading; flowers small, white; seed-vessel globose, many-seeded; filaments not toothed.

Three species of Scurvy Grass are alpine, viz.: *C. saxatilis*, Lam. (*Kernera saxatilis*, Rchb.); stem 6–8 in. high, diffusely branched, radical leaves obovate, dentate, or lyrate, common; *C. alpina*, Tausch.; stem

dwarf, 1–2 inches, radical leaves spathulate, usually entire; Tirol, rare; and *C. officinalis*, L. (*pyrenaica*, DC.); with cordate more fleshy radical leaves and fleshy stem; abundant on wet mountain-sides. *C. Armoracia*, L. (*Armoracia rusticana*, Wett.), Horse-radish, grows in ditches in the lowlands.

20. PETROCALLIS, Br.

Petals entire, lilac or pink; seed‑vessel oval, few-seeded; leaves in a dense rosette, 3-cleft at the apex.

P. pyrenaica, R.Br.; a pretty dwarf alpine plant with pink flowers and wedge-shaped leaves in dense tufts; common in the Alps, Dauphiny, and Pyrenees.

21. EROPHILA, DC.

Petals bifid, white; flowers small, on a leafless scape. Not alpine.

E. vulgaris, DC. (*Draba verna*, L.), Whitlow Grass; a very small plant, abundant on walls, &c., in the very early spring.

22. DRABA, L.

Petals entire, yellow or white; flowers usually crowded; filaments not toothed; radical leaves entire; seed-vessel many-seeded.

A. Flowers yellow:—*D. aizoides*, L. (Pl. 7); a beauti-ful alpine plant, with rosettes of rigid linear ciliated leaves and a leafless scape, 1–4 in. high, bearing a short corym-bose raceme of bright yellow flowers; abundant in the Alps, Jura, Pyrenees, and Dauphiny. *D. affinis*, Host.; alpine chain, local; differs in having the petals four times as long as the sepals (instead of twice). *D. Hoppeana*, Rchb. (*Zahlbruckneri*, Host.) ; Switzerland, Styria, rare ; has a much shorter style and a few leaves on the stem. *D. Sauteri*, Hoppe, and *D. Spitzelii*, Hoppe ; Styria, Tirol, Salzburg ; are dwarf forms, about 1 in. high, with short lanceolate ciliate leaves, the latter covered with spreading hairs. *D. cuspidata*, Bieb. ; Pyrenees ; scarcely differs from *D. aizoides* except in its broader seed-vessel. *D. nemorosa*, L. ; flowers large, stem-leaves few, with rounded base ; rocky places ; Pyrenees.

The differences between some of the species in the following sections are but slight, and hybridisation is probably not uncommon.

B. Flowers white ; radical leaves forming a single rosette :—*D. incana*, L. ; stem very leafy, petals twice as long as sepals ; rocky, local. *D. Thomasii*, Koch ; stem-leaves few, petals not much longer than sepals ; Switzerland, Tirol, Carinthia, local.

C. Flowers white ; radical leaves in numerous rosettes ; very cæspitose :—*D. pumila*, Miel. ; stem leafless, scarcely higher than the rosette of leaves ; Salzburg. *D. tomentosa*, Whlb. ; flowers large, petals emarginate, stem-leaves and leaf-stalk tormentose, leaves cordate, coarsely dentate, silicule ciliate ; dry, local. *D. frigida*, Saut. (*dubia*, Sut.) ; stem few-leaved, leaves grey-green on both sides, hairy, silicule glabrous ; dry, local. *D. Pacheri*, Stur. ; leaves yellowish-green above, entire or coarsely dentate, densely hairy beneath ; Southern Tirol, Carinthia, rare. *D. stellata*, Jacq. ; stem glabrous, stem-leaves dentate, petals with short claw, emarginate, style of silicule long ; Alps. *D. nivea*, Saut. ; stem glabrous, petals snow-white, with long claw, style of silicule very short ; Tirol, Carinthia, rare. *D. Wahlenbergii*, Hartm. (*lapponica*, DC.); stem and leaves nearly glabrous, silicule spreading, style very short; dry, frequent. *D. Johannis*, Host. (*carinthiaca*, Hoppe); leaves and stem with scattered stellate hairs, stem glabrous; dry, frequent. *D. muralis*; stem-leaves cordate, half amplexicaul, dentate; stony places in the lowlands.

D. Flowers white; leaves ciliate:—*D. ciliata*, Whlb.; leaves rather thick, shining, glabrous, petals three times as long as sepals, silicule linear; Tirol, Carinthia. *D. lactea*, Adans.; leaves ciliate at the base, stem stellate-hairy below, silicule elliptic-lanceolate; Tirol, Carinthia, rare.

23. LUNARIA, L.

Seed-vessel broadly elliptical, on a long stalk; two of the sepals gibbous at the base; petals large, violet.

L. rediviva, L., Moonflower; with large cordate-lanceolate, dentate leaves, and large fragrant flowers; in bushy places in the Alps, Dauphiny, Pyrenees.

24. CLYPEOLA, L.

Sepals equal at the base; filaments winged and toothed; silicule orbicular, winged, containing only a single seed. Not alpine.

C. Gaudini, Trachs.; flowers yellow, fruit-stalk arched; slopes; Southern Switzerland, rare.

25. VESICARIA, Poir.

Seed-vessel nearly globular, sessile, not winged.

V. utriculata, Poir., with lanceolate leaves and pale yellow flowers; clefts of rocks; Valais, Dauphiny.

Tribe CAMELINEÆ.—Silicule short, dehiscent, pyriform or turbinate; seeds in 2 rows.

26. CAMELINA, Crntz.

Flowers yellow; leaves auricled.   Not alpine.

C. sativa, Fr., Gold of Pleasure; *C. fœtida*, Fr.; and *C. sylvestris*, Fr.; are weeds, chiefly among flax, in Southern Switzerland and Pyrenees.

Tribe LEPIDINEÆ.—Seed-vessel short (silicule), dehiscent, and compressed at right angles to the septum; radicle incumbent; flowers white.   Genera 27–29.

27. CAPSELLA, Mœnch.

Valves of silicule boat-shaped, keeled; seeds numerous; flowers small.

C. bursa-pastoris, Mœnch., Shepherd's Purse; everywhere.   *C. pauciflora*, Koch; lower leaves 3-cleft, upper lanceolate, undivided, raceme 3–6 flowered; Switzerland, Tirol, rare.   *C. rubella*, Reut.; petals not longer than the sepals, which are streaked with red; Western Switzerland, rare.   *C. procumbens*, Fr.; racemes elongated, leaves pinnatifid; Fribourg.

28. SENEBIERA, DC.

Silicule didymous, 2-seeded; flowers very small.   Not alpine.

S. Coronopus, Poir.; silicule deeply wrinkled; a common weed.   *S. didyma;* silicules didymous; occasional.

29. LEPIDIUM, L.

Sepals very small; petals very small or 0; silicule oblong, 2–4-seeded.   Not alpine.

Several species of Cress are weeds in cultivated land, viz., the English species, *L. ruderale*, L., *campestre*, Br., and *Draba*, L.; also *L. graminifolium*, L., with linear-lanceolate stem-leaves; in Southern Switzerland and Pyrenees.

Tribe THLASPIDIEÆ.—Silicule short, dehiscent, compressed at right angles to the septum, horizontal; radicle accumbent; flowers white or lilac.   Genera 30–35.

30. THLASPI, L.

Silicule entire or notched; petals equal; sepals erect, equal; filaments without scales.

A. Flowers lilac, violet, or pink, rarely white; alpine plants:—*T. rotundifolium*, L.; silicule rounded at the apex, radical leaves roundish, stalked, stem-leaves amplexicaul; frequent; Switzerland, Dauphiny.   *T. corymbosum*, Gay; a more compact plant with lanceolate radical leaves; Zermatt.   *T. cepeæfolium*, Koch; silicule notched at the apex, radical leaves dentate; Carinthia, Carniola, Tirol.

B. Flowers white; the following are alpine plants:— *T. alpestre*, L.; with the raceme elongated in fruit, the silicule broadly winged and notched, and leaves nearly entire; Switzerland, Dauphiny, Pyrenees.   *T. Salisii*, Brügg.; resembling the last, but with a more branched stem and dentate leaves; Switzerland, Tirol.   *T. alpinum*, Crntz.; growing in loose tufts, the silicule narrowly winged and scarcely notched, stem-leaves cordate; pastures; generally distributed.   *T. Mureti*, Gremli; resembling the last, but with longer stamens.   *T. montanum*, L.; with larger flowers and obcordate broadly winged silicule; Pyrenees, Dauphiny, Jura; and *T. præcox*, Wulf. (probably only a variety of the last); Tirol, Styria, Carniola. *T. perfoliatum*, L., with obcordate silicule and cordate amplexicaul stem-leaves, is found in lowland calcareous districts; and *T. arvense*, L., Penny Cress, with broadly winged silicule and sagittate stem-leaves, is a weed in cultivated land.

31. IBERIS, L.

Sepals equal; petals very unequal, the two outer ones larger; flowers white or pink; silicule 2-seeded.

I. amara, L., Candytuft; flowers usually white; cultivated land. Also the following in rocky places in the Pyrenees:—*I. spathulata ;* leaves nearly orbicular, flowers lilac. *I. Bernardiana*, G. and G. (Pl. 8), (*I. nana*, Lap.); sepals and petals violet, leaves linear-oblong. *I. ciliata*, All.; flowers white or slightly purple, leaves linear, ciliated. *I. Garrexiana*, All., Perennial Candytuft; flowers white, leaves thick, linear-obovate, flower-stalk thickened in fruit. *I. saxatilis ;* leaves fleshy, mucronate, flowers white, sepals coloured at the edge; also in Jura, Dauphiny. *I. pinnata*, L., with pinnate, and *I. panduræformis*, Pourr., with deeply toothed stem-leaves, occur occasionally in cultivated ground.

32. TEESDALIA, Br.

Flowers small, white; petals usually unequal; filaments with a scale at the base. Not alpine.

T. nudicaulis, Br.; stem 4–6 in. high, leafless; two outer petals twice as long as inner ones; sandy and gravelly places.

VIII.—IBERIS BERNARDIANA.

33. HUTCHINSIA, Br.

Flowers small, white; petals equal; filaments without scales; silicule 2–16-seeded; leaves pinnatifid.

H. alpina, Br.; stem leafless, 1–3 in. high, silicule 4-seeded; Jura, Dauphiny, Pyrenees. *H. brevicaulis*, Br.; a smaller plant, with shorter fruit-raceme, silicule 2-seeded; moist alpine rocks. *H. procumbens*, Desv.; stem leafy, prostrate, silicule 12–16-seeded; Pyrenees, Southern Switzerland. *H. petræa*, Br.; stem ascending, leafy, petals not longer than sepals, silicule 4-seeded; limestone rocks.

34. ÆTHIONEMA, Br.

Sepals unequal, two gibbous at the base; longer filaments toothed; silicule winged; valves boat-shaped.

A. saxatile, Br.; leaves thick, glabrous, flowers very small, pink or white; rocky; Pyrenees, Southern Switzerland, Dauphiny.

35. BISCUTELLA, L.

Sepals usually equal; flowers yellow; stem leafy; silicule separating into two distinct 1-seeded valves.

B. lævigata, L.; stem erect, much branched, leaves shining, upper ones amplexicaul, flowers fragrant; rocky; Alps, Pyrenees, Dauphiny. *B. cichoriifolia*, Lois.; two of the sepals gibbous; Pyrenees, Carniola.

Tribe ISATIDEÆ. — Silicule indehiscent, unilocular, 1-seeded. Genera 36–38.

36. ISATIS, L.

Petals equal; flowers yellow; silicule compressed, flat.

I. tinctoria, L., Dyer's Woad; stem erect, branched, stem-leaves arrow-shaped; cultivated land in the South.
I. alpina, All.; a smaller more leafy plant, flowers larger, bright yellow; pastures; Pyrenees, Dauphiny.

37. NESLIA, Desv.

Petals equal; flowers yellow; silicule globular.

N. paniculata, Desv.; stem slender, branched, leaves entire or slightly toothed, silicule wrinkled, flowers small, pale yellow; pastures; Jura, Pyrenees.

38. CALEPINA, Adans.

Petals unequal; silicule globular, beaked.

C. Corvini, Desv.; flowers small, white, the two outer petals larger than the two inner ones, radical leaves runcinate, stem-leaves auricled, arrow-shaped; grassy places; Southern Switzerland, Pyrenees.

Tribe BUNIADEÆ. — Silicule indehiscent, with 2–4 1-seeded loculi.

39. BUNIAS, Br.

Petals equal, entire or emarginate; silicule 4-angled. Not alpine.

B. Erucago, L.; flowers small, yellow, radical leaves runcinate, stem-leaves lanceolate, sessile; fields; Western and Southern Switzerland.

Tribe RAPHANEÆ.—Silicule indehiscent, jointed.

40. RAPHANUS, L.

Silicule elongated, separating into several 1-seeded joints; flowers white or pale pink. Not alpine.

R. Raphanistrum, L., Wild Radish; cultivated land.

41. RAPISTRUM, Boerh.

Silicule separating into two 1-seeded joints; flowers yellow. Not alpine.

R. rugosum, Bergt.; lower leaves lyrate, upper oblong, sessile; road-sides; Western Switzerland, Pyrenees. *R. perenne*, All.; all the leaves pinnatifid, acute; Southern Switzerland.

Order VII.—RESEDACEÆ.

Flowers in racemes, irregular; sepals unequal; petals unequal, deeply divided, springing from a broad nectariferous disk; stamens 10–40, pendant.

1. RESEDA, L.

Petals 4–7; seed-vessel a capsule, opening at the apex before the seeds are ripe. A very small order, chiefly Mediterranean.

The two British species, *R. lutea*, L., Wild Mignonette, and *R. Luteola*, L., Weld, are frequent; also *R. Phyteuma*, L.; sepals and petals 6, stem-leaves trifid at the apex, capsule large, subtended by the very large calyx; Pyrenees, Jura, Carniola; and *R. glauca*, L.; sepals and petals 5, leaves linear, entire, whole plant glaucous; sub-alpine valleys in the Pyrenees.

2. ASTROCARPUS, Neck.

Carpels 4–6, distinct, 1-seeded, each opening by the ventral suture.

A. sesamoides, Gay; flowers white, stamens 7–9, carpels 5, leaves linear-lanceolate, the radical ones forming a rosette; alpine and sub-alpine localities in the Pyrenees.

Order VIII.—CISTACEÆ.

Flowers conspicuous, fugacious; sepals usually 3; petals usually 5; stamens numerous; ovary usually 1-celled with three parietal placentæ; style 1; stigmas 3. A small order, chiefly Mediterranean.

1. CISTUS, Tourn.

Sepals 3–5; petals large, brightly coloured. Balsamic shrubs, with large handsome very fugacious flowers.

C. salvifolius, L.; flowers solitary, axillary, 1½–2 in., white with yellow base; Southern Switzerland, Locarno, Ascona, Pyrenees. Several other species are natives of the departments of the Pyrenees, but belong to the Mediterranean flora.

2. HELIANTHEMUM, Tourn.

Sepals 5, 2 of them usually smaller; leaves opposite.

A. Leaves stipulate :—*H. salicifolium*, Pers.; flowers opposite to the leaves, sepals ovate-lanceolate, not acuminate; Southern Switzerland (rare), Pyrenees. *H. vulgare*, Gærtn., our English Rock Rose; very abundant on dry

IX.—HELIANTHEMUM ROSEUM.

hill-sides; flowers yellow, occasionally rose-coloured (var. *roseum*, Pl. 9), especially in the Pyrenees, or white. *H. polifolium*, Pers.; flowers white, sepals tomentose, leaves hoary on both sides; Ticino, Jura, rare.

B. Leaves without stipules:— *H. alpestre*, Rchb. (*œlandicum*, DC.); flowers yellow, leaves slightly hairy; high elevations; Alps, Ticino, and Pyrenees. *H. canum*, Dun. (*vineale*, Pers.); flowers small, yellow, calyx hairy, leaves hoary beneath; at lower elevations; Jura, Dauphiny, Pyrenees. *H. Fumana*, Mill. (*Fumana procumbens*, G. and G.); capsule partially 3-celled, stamens 20–40, the outer ones sterile; Jura, Ticino, Dauphiny, Pyrenees. Several other species occur in the Pyrenean valleys.

Order IX.—VIOLACEÆ.

Leaves stipulate, axillary and solitary, or in small cymes, with two small bracts; calyx and corolla usually irregular; seed-vessel a 1-celled 3-valved capsule, with three parietal placentæ. A rather large order, distributed through the temperate and tropical regions of the globe.

1. VIOLA, L.

Flowers solitary or in pairs; sepals unequal; petals unequal, the lower ones larger and spurred at the base; anthers sessile, the two lower ones often with appendages projecting into the spur of the corolla.

A. Stipules not leafy; upper petals directed forwards; stigma beaked. The species of this section commonly produce closed apetalous flowers in the autumn, which are fertile. It includes the following familiar English species:

—*V. odorata*, L., Sweet Violet; *V. canina*, L., Dog Violet, with its numerous sub-species; *V. sylvatica*, Fr., Wood Violet, with its sub-species; *V. palustris*, L., with small pale blue flowers and reniform leaves, in marshes; and *V. hirta*, L., resembling *odorata*, but smaller and scentless, on open hill-sides. *V. Beraudii*, Bor., Rhone valley, is probably a variety of *odorata*, with blue flowers and a white throat; *V. multicaulis*, also of *odorata*, with dusky violet flowers; *perplexa*, Grml., a hybrid; *V. stagnina*, Kit. (*lactea*, Rchb.), and *elatior*, Fr., both occasional in swampy meadows, varieties of *canina*.

The following species are mostly alpine or subalpine :—

a. Stigma terminating in an oblique disk; seed-vessel pendant :—*V. pinnata*, L.; leaves all radical, palmate with pinnatifid segments, stem-leaves o, flowers light violet, fragrant; very high; Switzerland, Tirol, Dauphiny, Carinthia, blossoming very early.

b. Style pointed, curved; fruit-stalk horizontal on the ground; seed-vessel globular :—*V. alba*, Bess.; flowers white, fragrant, sepals obtuse, leaves ovate-acuminate, stipules very narrow, seed-vessel hairy; Switzerland, very rare; Jura, Zug. *V. sciaphila*, Koch; flowers small, violet, fragrant, sepals oval, petals bearded, seed-vessel glabrous; Switzerland, Tirol, Dauphiny, Carinthia, rare. *V. ambigua*, Koch (*Thomasiana*, P. and S.); leaves lanceolate-ovate, flowers large, violet-red, fragrant, seed-vessel hairy; Ticino, Pusterthal. *V. collina*, Bess.; flowers blue, fragrant, leaves cordate-ovate, seed-vessel globular, pubescent, stipules fimbriate, pubescent; Switzerland, lowlands.

c. Style pointed, curved; fruit-stalk erect; seed-vessel

trigonous :— *V. arenaria*, DC.; flowers pale blue, stipules fimbriate, whole plant covered with a soft down ; Switzerland, Dauphiny, rare. *V. mirabilis*, L.; flowers pale blue, fragrant, stipules ovate-lanceolate, ciliate, sepals ovate-lanceolate, acute ; Switzerland, rare. *V. pratensis*, Koch (*pumila*, Vill.) ; flowers blue, leaves decurrent on the petiole, stipules inciso-dentate ; damp places ; Jura, Southern Switzerland, rare. *V. uliginosa*, Bess.; stem very short, spur short, leaves obovate, sub-cordate, stipules entire ; moist places ; Carniola, Carinthia.

B. Stipules not leafy ; style thickened ; stigma bifid ; the two upper and two lateral petals directed upwards : —*V. biflora*, L. (Pl. 10); flowers yellow streaked with brown, two in each leaf-axil, leaves kidney-shaped, crenate ; Alps, Jura, Pyrenees, frequent.

C. Style pointed, curved; stipules leafy :— *V. montana*, L. (*stricta*, Horn.); flowers large, violet-blue, sepals acute, spur obtuse, green, stipules dentate ; Southern Switzerland, Tirol. *V. Schultzii*, Bill.; flowers pale yellow then white, spur green then yellow, leaves cordate, stipules dentate ; Jura, Carniola, Tirol.

D. Stipules leafy ; the two upper and two lateral petals directed upwards ; stigma globular, hollow, bearded :— *V. tricolor*, L., Heartsease, Pansy, with its sub-species, in cultivated land ; also the following alpine and subalpine species :— *V. alpina*, L.; flowers large, blue, rarely white, radical leaves rounded-ovate, crenate, stipules adnate to the leaf-stalk, stem-leaves 0 ; Eastern Switzerland, Tirol, Carpathians, frequent. *V. cenisia*, L., flowers violet, leaves entire, hispid, spur of corolla as long as calyx, stipules entire ; Switzerland, Tirol, Dauphiny, Pyrenees. *V. Comollia*, Mass.; resembling

the last, but spur shorter than calyx, sepals fringed;
Southern Tirol. *V. calcarata*, L.; flowers large, violet
or yellow, fragrant, spur slender, as long as the corolla,
leaves ovate-lanceolate, crenate, stipules spathulate,
nearly entire; Jura, Tirol, Carinthia. *V. lutea*, Huds.;
flowers yellow or variegated with blue, lower leaves
cordate-ovate, crenate, upper lanceolate, stipules very
deeply divided; marshy ground, frequent. *V. heterophylla*, Bert.; resembling the last, but with a longer
spur, flowers blue; dry, calcareous; Southern Tirol.
V. cornuta, L.; flowers blue, sepals with very long
appendages, spur of corolla elongated, leaves ovate,
crenate, stipules dentate; Pyrenees. *V. nummulari-
folia*, All.; flowers small, violet, spur short, leaves
rounded, entire; Dauphiny.

Order X.—POLYGALACEÆ.

Flowers irregular; 2 inner sepals large, petaloid (wing-
sepals); stamens 8; ovary 2-celled, 2-seeded. A rather
large order, chiefly tropical and sub-tropical, represented
in Europe by a single genus.

1. POLYGALA, L.

Petals combined below with the split sheath composed
of the eight filaments; corolla usually crested; anthers
opening by oblique pores; stigma spathulate.

A. Corolla not crested; filaments distinct nearly to
the base: —*P. Chamæbuxus*, L. (Pl. 11); suffruticose,
flowers large, corolla usually yellow, wing-sepals
purple or white, leaves ovate, coriaceous, evergreen;

XI.—POLYGALA CHAMAEBUXUS.

XII.—POLYGALA CALCAREA.

a beautiful plant, frequent in bushy places in the Alps, Dauphiny, and Pyrenees.

B. Corolla with a fimbriate crest; filaments united half-way; flowers much smaller :—*P. vulgaris*, L., Milkwort; flowers blue, pink, or white, very common on dry banks. *P. depressa*, Wend. (*P. serpyllacea*, Weihe); stem more procumbent, flexuous; *P. calcarea*, Schultz (Pl. 12); a much more compact plant, with spreading leafy branches; *P. austriaca*, Crntz. (*uliginosa*, Rchb.); with small flowers and very narrow wing-sepals; *P. amara*, L. (*P. alpestris*, Rchb.); a smaller plant with small flowers, nearly simple stem, thick spathulate leaves, and narrow inner sepals; all with blue flowers; in similar localities. *P. comosa*, L.; resembling *P. vulgaris*, but a larger erect plant, with the bracts greatly exceeding the unopened flowers; hillsides; Southern Switzerland, Tirol, Pyrenees. *P. rosea*, Desf. (*nicæensis*, Riss.); flowers large, pink, in terminal spikes, wing-sepals very large, $\frac{1}{4}$–$\frac{1}{2}$ in., nearly round, capsule with a broad border; Ticino, Simplon, Pyrenees.

Order XI.—DROSERACEÆ.

Sepals 4–8; petals 4–8; stamens 4–20; ovary free, 1–5-celled; styles 1–5. A small order of aquatic or marsh plants with insectivorous habit.

1. DROSERA, L.

Flowers small, white, in scorpioid cymes, on leafless scapes; radical leaves fringed with capitate tentacles, which exude a mucilaginous excretion. The pretty little

Sundews are remarkable for their insectivorous habit, insects being detained by the viscid excretion from the stalked glands which cover the margin and upper surface of the leaf, and then digested in a chamber formed by the infolding of the leaf.

The three British species, *D. rotundifolia*, L.; *intermedia*, Hay.; and the larger and less common, *D. anglica*, Huds. (*longifolia*, L.); are found in similar situations throughout the region, in *Sphagnum* bogs and by mountain streams.

2. ALDROVANDA, L.

Flowers solitary, axillary; petals 5; sepals 5; leaves in whorls.   Aquatic.

A. vesiculosa, L.; the whorls of 6–9 very thin submerged leaves bear bladders, in which small aquatic animals are captured and digested; a Southern plant; Lake of Constance, Tirol, Pyrenees.

Order XII.—CARYOPHYLLEÆ.

Flowers regular; inflorescence always cymose; sepals and petals usually 5 each; stamens usually 10, in two rows; ovary 1-celled with central placentation, or partially 2–5-celled with axile placentation; leaves always opposite and entire, springing from swollen joints. A large order, almost confined to the arctic and temperate zones of the Old World.

Tribe SILENEÆ.—Sepals 4–5, more or less united at the base; styles 2–5, free; disk elongated, bearing the petals and stamens.   Genera 1–8.

1. GYPSOPHILA, L.

Calyx campanulate, 5-angled, without scales at the base; petals not clawed, gradually narrowed to the base, without a corona; styles 2.

G. muralis, L.; an annual plant, with somewhat pubescent stem, flowers inclined, petals white, veined with red; on walls in the lowlands. *G. repens*, L.; a perennial very glabrous plant, inflorescence many-flowered, flowers erect, petals usually white, sometimes pink; in all alpine situations.

2. TUNICA, L.

Calyx campanulate, 5-angled, with 2 or more scales at the base; otherwise like *Gypsophila*.

T. saxifraga, L.; dry places in all alpine regions, common.

3. DIANTHUS, L.

Calyx tubular-campanulate, with two or more imbricating scales at the base, 5-toothed; petals 5, clawed, usually cut or fringed, pink or red; ovary 1-celled; styles 2.

Many species of Pink are alpine plants, growing at a great elevation; others belong to the Mediterranean flora.

A. Flowers in crowded terminal cymes enclosed in a common involucre:—*D. barbatus*, L. (Sweet William); leaves broadly lanceolate, shortly stalked, sheathing, calyx-scales green, petals irregularly dentate; Switzerland (rare), Pyrenees, Carniola, Carinthia, Styria.

D. Armeria, L., Deptford Pink, flowers small ($\frac{1}{2}$ in. diam.), in loose cymes, petals narrow, distant, toothed, red with dark spots, bracts green, downy; dry banks in the lowlands. *D. prolifer*, L.; leaves short with scabrous margins, petals contiguous, notched, purplish red; dry pastures, common. *D. Carthusianorum*, L. (Pl. 13); leaves linear-lanceolate, acute, calyx ciliate, petals contiguous, irregularly dentate, uniformly red, involucral bracts brown, coriaceous; dry meadows, common. *D. atrorubens*, All. (*vaginatus*, Vill.); similar, but with smaller darker flowers, petals not contiguous, and slender stem; Southern Switzerland, Tirol, Dauphiny, Pyrenees.

B. Flowers solitary or in loose cymes; petals toothed: —*D. Seguieri*, Chx. (Pl. 14); flowers large, usually 2–4, petals deeply toothed, pink with a row of red spots at the base, bearded, contiguous, calyx cylindrical, calyx-scales ciliate; alpine; Southern Switzerland, Dauphiny, Pyrenees. *D. attenuatus*, Sm.; flowers smaller, usually 2–4, uniformly pink, calyx conical, petals not contiguous, crenulate or dentate, leaves subulate; rocks; Eastern Pyrenees. *D. pungens*, L.; flowers 1–3, pink, not contiguous, nearly entire, leaves subulate, glaucous, tufted; rocks; Eastern Pyrenees. *D. subacaulis*, Vill.; flowers solitary, very small, pink, petals not contiguous, nearly entire, calyx cylindrical, leaves stiff; a very dwarf plant, forming a glaucous turf; Dauphiny. *D. neglectus*, Lois.; flowers usually solitary, purple, petals not contiguous, deeply toothed, outer sepals with a short stiff awn, leaves 3-nerved; Mont Cenis, Dauphiny, Pyrenees. *D. deltoides*, L., Maiden Pink; flowers solitary, $\frac{3}{4}$ in. diam., pink spotted with white, petals not contiguous, leaves soft, downy, 3-nerved; mountain pastures, common.

XIV.—DIANTHUS SEGUIERI.

D. cæsius, Sm., Cheddar Pink ; flowers usually solitary, pink, 1 in. diam., fragrant, petals contiguous, bearded, calyx cylindrical; Switzerland, Tirol. *D. sylvestris*, Wulf.; flowers solitary, peach-coloured, petals contiguous, not bearded, nodes of stem purple, leaves soft, flat (stem often wanting, plant densely cæspitose, *D. frigidus*, Koch); high elevations. *D. alpinus*, L.; flowers solitary, large, fragrant, petals crimson spotted with white, greenish on the under side, jagged, stem 2–3 in. high, leaves linear-lanceolate, 1-nerved, outer sepals with a membranous awn; high alpine pastures; Switzerland, Tirol, Carpathians. *D. glacialis*, Hänke (Pl. 15); resembling the last, but petals not spotted, leaves linear; very high ; Switzerland, Tirol, Styria, Carinthia.

C. Flowers solitary or in loose cymes; petals fringed or slit :—*D. superbus*, L. (*speciosus*, Rchb.); flowers large, solitary, very deeply fringed, pink, very fragrant, leaves linear-lanceolate; moist pastures; Jura, Southern Tirol, Dauphiny, Pyrenees. *D. monspessulanus*, L. (*alpestris*, S. and H.); resembling the last, flowers lighter, petals not so deeply divided, leaves linear, acuminate; Ticino, Jura, Dauphiny, Pyrenees, Carniola, Carinthia, Styria. *D. gallicus*, Pers. (Pl. 16); flowers solitary, pale pink or white, very fragrant, petals slit to one-third of their length, calyx cylindrical, whole plant pale green, glaucous; sandy; Western Pyrenees. *D. plumarius*, L.; flowers solitary, large, 1 in. diam., pink, deeply cut, fragrant, leaves grooved above, 1-nerved; Styria, Carniola, Carinthia : said to be the origin of the garden Pinks. *D. Caryophyllus*, L., Carnation, Clove Pink, is not wild in Switzerland.

4. SAPONARIA, L.

Flowers in compound cymes; calyx tubular; petals 5, clawed; seed-vessel bursting by 4 valves; styles 2; disk small.

S. officinalis, L., Soapwort; fields and wet places. *S. Vaccaria*, L.; stem ascending, 15–20 in., flowers pink, leaves lanceolate, acute, the upper ones somewhat amplexicaul; cultivated land. The following are alpine:— *S. ocymoides*, L. (Pl. 17); stem decumbent, petals pink or lilac, leaves broadly lanceolate; Switzerland, Tirol, Carinthia, Pyrenees. *S. lutea*, L.; flowers crowded, petals yellow, violet at the base, stem erect, simple, leaves linear; very rare; Monte Rosa, Dauphiny. *S. cæspitosa*, DC.; plant cæspitose, flowers pink, leaves linear, coriaceous, keeled; higher Pyrenees.

5. SILENE, L.

Calyx inflated, 5-toothed; petals 5, clawed, with 2 scales at the base of the blade, forming a corona; disk columnar; ovary partially 3-celled; styles 3; seed-vessel bursting into 6 valves.

The following lowland species are English, all with white flowers:—*S. inflata*, Sm. (*S. Cucubalus*, Wib.), Catchfly, Bladder Campion, common; *S. nutans*, L., Nottingham Catchfly, frequent; *S. Otites*, L., Switzerland; *S. conica*, L., Dauphiny, Pyrenees; *S. gallica*, L., Dauphiny, Pyrenees, Jura; *S. noctiflora*, L. (*Melandrium noctiflorum*, Fr.), flowers fragrant at night, frequent. Also in the southern districts:—*S. Armeria*, L., with small pink flowers, rarely white, and broad ovate leaves;

XVI.—DIANTHUS GALLICUS.

XVII —SAPONARIA OCYMOIDES.

S. italica, Pers., with small fragrant white or light pink flowers, hairy stem, and narrow ciliated leaves; and *S. linicola*, Grml., a weed among flax.

The following are alpine species :—

A. Flowers white or slightly coloured :—*S. alpina*, Thom. (*inflata*, var. *maritima*, With.), cæspitose, flowers large, calyx swollen; closely resembling *S. inflata*, but with darker anthers and more fleshy leaves; common at high altitudes. *S. alpestris*, Jacq.; stem somewhat viscid, branched, petals 4-toothed, calyx glandular-hairy, leaves lanceolate, coriaceous; Switzerland, Tirol, Carniola, frequent. *S. quadrifida*, L.; petals 4-toothed, stem viscid above, leaves linear; Switzerland, Pyrenees, frequent. *S. rupestris*, L.; petals notched, milk-white, sometimes reddish, twice as long as calyx, leaves broadly lanceolate; frequent. *S. saxifraga*, L.; stem 4–6 in. high; petals deeply bifid, greenish or reddish beneath, calyx glabrous; Switzerland, Dauphiny, Pyrenees, rare. *S. paradoxa*, L.; inflorescence pyramidal, flowers nocturnal and fragrant, petals bipartite, stem viscid above; Dauphiny.

B. Flowers pink or red, rarely white :—*S. Pumilio*, Wulf.; flowers large, solitary, petals undivided, calyx elongate-campanulate, reddish, hairy; Eastern Switzerland, at high altitudes. *S. Elisabethæ*, Jan.; flowers few, large, stem simple, downy, erect, petals obcordate, deeply bilobed, lobes sharply toothed, leaves broadly lanceolate; Tirol, rare. *S. pudibunda*, Hoff.; petals 4-toothed, overlapping, leaves broadly linear; Switzerland, rare. *S. valesia*, L.; flowers solitary or very few, large, petals bifid, calyx very elongated, pubescent; rare; Monte Rosa, Great St. Bernard, Dauphiny.

S. acaulis, L., Moss Campion (Pl. 18); densely cæspitose, flowers small, solitary, calyx campanulate, seedvessel oblong, longer than calyx; one of the prettiest of the common alpine plants. *S. exscapa*, All.; resembling the last, but petals light pink, seed-vessel oval, scarcely longer than corolla; Switzerland, local. *S. ciliata*, Pourr.; flowers solitary or in a unilateral cyme, petals bifid, leaves soft, hairy, ciliate; Pyrenees.

6. Cucubalus, Gært.

Petals 5, with scales at the throat forming a corona; stamens 10; styles 3; fruit a berry. Not alpine.

C. bacciferus, L.; flowers pale green, petals deeply bifid, berry black when ripe, subtended by the cup-like calyx; damp places in the lowlands.

7. Lychnis, L.

Flowers often diœcious; petals 5, divided, with one or two scales at the base of the lamina, red or white; styles 5.

L. diurna, Sibth. (*Melandrium diurnum*, Crep.), Red Campion; *vespertina*, Sibth.; and *Flos-cuculi*, Ragged Robin, are common English plants. *L. Viscaria*, L. (*Viscaria vulgaris*, Rœhl), (Pl. 19); stem viscid at the nodes, flowers crowded, calyx swollen, petals pink, notched, lower leaves lanceolate, upper narrower; dry hill-sides; sub-alpine. *L. Flos-Jovis*, Lam. (*Agrostemma Flos-Jovis*, L.); flowers crowded, petals deeply divided, pink, leaves thick, lanceolate-acuminate, stem and leaves covered with a white down; exposed sub-alpine pastures;

XVIII.—SILENE ACAULIS.

Southern Switzerland, Savoy, Dauphiny. *L. coronaria*, Lam. (*Agrostemma coronaria*, L.); flowers large, in a loose cyme, petals pink, entire, leaves thick, lanceolate, stem and leaves tomentose; stony places; Southern Switzerland, Pyrenees, rare. *L. alpina*, L. (Pl. 20); flowers in compact cymes, petals pink, bifid; lower leaves lanceolate, upper narrower, stem 3–6 in. high; a cæspitose glabrous plant; high altitudes; Switzerland, Dauphiny, Pyrenees. *L. nummularia*, Lap. (*Petrocoptis pyrenaica*, Br.); cæspitose, flowers white, in loose cymes, calyx campanulate, petals entire or emarginate, wedge-shaped, styles 5–6; rocks; Western Pyrenees.

8. GITHAGO, Desf.

Calyx coriaceous, with leafy teeth; petals entire, destitute of a corona, clawed; styles 5. Not alpine.

G. segetum, Desf. (*Agrostemma Githago*, L.), Corn Cockle; woolly, flowers large, solitary, petals light purple; cultivated land.

Tribe ALSINEÆ.—Sepals distinct; petals clawed, not furnished with a corona, usually white and small; disk small; styles free. Genera 9–21.

9. HOLOSTEUM, L.

Petals jagged; styles 3; capsule cylindrical, 6-valved; stem viscid. Not alpine.

H. umbellatum, L.; stamens often only 3, radical leaves narrowly elliptic, stem nearly leafless; walls and dry places; not common.

10. CERASTIUM, L.

Flowers white, usually small; petals notched or bifid; styles 3 or 5; capsule cylindrical, curved, 6- or 10-toothed, when ripe usually greatly exceeding the calyx. Herbs with pubescent stem and leaves, mostly prostrate.

The following English lowland species of Mouse-ear Chickweed are found also in Switzerland:—*C. glomeratum*, Fr.; *semidecandrum*, L.; *triviale*, Lk.; *arvense*, L.; *vulgatum*, L.; *viscosum*, L. *C. brachycarpum*, Schm., and *suffruticosum*, L., are probably mountain forms of *arvense*; and *C. macrocarpum*, Schm., of *vulgatum*. *C. brachypetalum*, Pers., is very nearly allied to *glomeratum*. *C. glutinosum*, L., covered with a glutinous down, is a Southern lowland species.

The following are more or less alpine, but the specific characters are often very difficult to determine:—*C. grandiflorum*, W. and K.; plant covered with a thick grey tomentum, leaves linear, fleshy, teeth of capsule revolute; very rare; Upper Styria. *C. lanatum*, Koch; resembling the last, but with broader leaves, teeth of capsule straight; very high; Southern Switzerland, rare. *C. alpinum*, L.; usually more or less glandular-hairy, stem 1–5-flowered, with rosettes of leaves, flower-stalk oblique after flowering, sepals obtuse, with a membranous margin; high altitudes; frequent. *C. latifolium*, L.; leaves ovate-elliptical, stiff, brittle, flowers large, few, petals more than twice as long as sepals, deeply bifid, capsule nearly globose; high; Switzerland, Dauphiny. *C. uniflorum*, Mur.; resembling the last, but leaves soft, very hairy, capsule narrower; Switzerland. *C. pyrenaicum*, Gay; similar, but petals and stamens ciliate,

XX.—LYCHNIS ALPINA.

flowers smaller; Eastern Pyrenees. *C. filiforme*, Schleich.; corolla bell-shaped, plant very slender, with long flower-stalks; high glaciers in the Alps. *C. ovatum*, Hoppe; stem cæspitose, 6–9 flowered, sepals with a broad white margin, bracts with a broad membranous margin; dry rocks, frequent. *C. alpicolum*, Brügg.; stem densely cæspitose, 1–3 in. high, 5–10-flowered, stem and leaves glandular-hairy; rocks at a high elevation. *C. trigynum*, Vill. (*Stellaria cerastoides*, L.); nearly glabrous, petals deeply bifid, styles 3, leaves glabrous, often recurved; moist alpine situations, frequent.

11. MŒNCHIA, Ehrh.

Flowers small, white; petals entire, shorter than the calyx; stamens 4–10. Small glabrous herbs; not alpine.

M. erecta, Ehrh. (*Cerastium quaternellum*, Fnzl., *C. glaucum*, Gren.); sepals, petals, styles, and stamens 4 or 8 (*octandrum*, Gay), flowers on long stalks, branches slender, stiff, whole plant glaucous; sandy places in the lowlands. *M. mantica*, Bartl.; sepals, petals, and styles 5, stamens 10; Southern Switzerland, Carniola, rare.

12. STELLARIA, L.

Flowers white, in dichotomous cymes; petals 5, bifid, without a corona, clawed; stamens 10; styles 3; capsule splitting into 6 valves. Not alpine.

The following British species occur:—*S. media*, Vill., Chickweed; everywhere. *S. Holostea*, L., Larger Stitchwort; and *graminea*, L., Smaller Stitchwort; hedges and grassy places. *S. nemorum*, L.; woods. *S. palustris*,

Ehrh. (*glauca*, With.); and *uliginosa*, Murr.; wet places, the latter ascending to high elevations. *S. Friesiana*, Ser. (Engadine), resembles *uliginosa*, with longer petals and narrower leaves. *S. pallida*, Piré, is an apetalous variety of *media*.

13. MALACHIUM, Fr.

Resembling *Stellaria*, but styles 5; capsule splitting into 5 bifid valves. Not alpine.

M. aquaticum, Fr. (*Stellaria aquatica*, Scop.); wet places.

14. MŒHRINGIA, L.

Petals 5, rarely 4, expanded; stamens 10, rarely 8; styles usually 3; capsule 3–6-valved; seeds with a mantle-like appendage at the base. Cæspitose, mostly alpine plants, with small white flowers.

A. Sepals and petals 4; stamens 8; styles 2:—*M. muscosa*, L.; petals longer than sepals, leaves linear, plant very fragile; high moist places, common.

B. Sepals and petals 5; stamens 10; styles 3:—*M. sphagnoides*, Rchb.; stem densely cæspitose, flowers on very short stalks, leaves imbricate, trigonous; Tirol, rare. *M. stenopetala*, Hausm.; petals not longer than sepals, very narrow; Tirol, very rare (Gross Glockner). *M. diversifolia*, Döll.; petals as broad as sepals, but not longer, lowermost leaves ovate, on long stalks, uppermost linear; wet rocks; Styria, Carniola, Carinthia, rare. *M. glaucovirens*, Bert. (*glauca*, Leyb.); leaves all linear, semi-cylindrical, glabrous, flowers on long stalks, petals not longer than sepals; Tirol, rare. *M. villosa*, Fnzl.;

petals longer than sepals, lanceolate, acute, nerveless, lowermost leaves elliptical; Carniola, rare. *M. polygonoides*, M.K. (*ciliata*, Scop.); petals broadly lanceolate, obtuse, 3-nerved, sepals obtuse, leaves linear, crowded; frequent on wet rocks. *M. bavarica*, L. (*Ponæ*, Fnzl., *dasyphylla*, Brun.); very fragile, glaucous, petals longer than sepals, leaves distant, apiculate, fleshy, stem 2–6 in. long; wet rocks; Tirol, Styria, Pyrenees. *M. trinervia*, Clairv.; petals shorter than sepals, leaves 3–5-nerved, ciliate; a common lowland plant.

15. ARENARIA, L.

Petals 4–5; stamens 8–10; styles usually 3; capsule 6-lobed; seeds without an appendage. Most species of Sandwort are alpine plants with small white flowers.

A. Petals twice as long as sepals; leaves lanceolate-subulate:—*A. grandiflora*, L.; leaves with a thickened margin, 1-nerved beneath; Jura, Tirol, Pyrenees.

B. Leaves ovate or ovate-lanceolate; petals rather longer than sepals:—*A. biflora*, L.; leaves nearly orbicular, blunt, petals elliptical; moist rocks, high, local. *A. Huteri*, Kern.; leaves ovate-lanceolate, acute, hairy, petals ovate; Southern Tirol. *A. ciliata*, L.; leaves ovate, ciliate; dry, rocky; Switzerland, Dauphiny, Pyrenees. *A. multicaulis*, L., is a more procumbent variety.

C. Leaves ovate or ovate-lanceolate; petals shorter than sepals:—*A. alpina*, Gaud. (*Marschlinsii*, Koch); petals ovate, sepals broadly lanceolate, acuminate, with a scarious margin; very high; Switzerland, Tirol, Carinthia. *A. Moritzii*, Brügg.; resembling the last, but petals elliptical, often tinged with red; rare; Engadine. *A. serpyllifolia*, L., is a very common plant in the lowlands; and *A. leptoclados*, Guss., a more delicate plant, with smaller flowers and narrower sepals, in the South.

D. Petals longer than sepals, usually tinged with red; capsule cylindrical :— *A. purpurascens*, Ram.; leaves ovate-lanceolate, glabrous or ciliate; Pyrenees, abundant.

16. ALSINE, Wahl.

Petals 5, rarely 4, entire or slightly notched; stamens 5, 8, or 10; styles usually 3; capsule 3-valved. Cæspitose or procumbent plants, usually with linear or linear-lanceolate leaves; mostly alpine.

A. Leaves roundish, ovate, ovate-lanceolate, or lanceolate :—*A. aretioides*, M.K. (*cherlerioides*, Schrad., *herniarioides*, Riou); plant forming a dense tuft, sepals and petals 4, stamens 8; very high; Switzerland. *A. lanceolata*, M.K. (*Facchinia lanceolata*, Rchb., *Stellaria rupestris*, Scop.), densely cæspitose, sepals and petals 5, stamens 10; high; Switzerland, Tirol, Dauphiny, Carinthia. *A. cerastifolia*, Fnzl.; flower-stalk 2-flowered, corolla often light pink; Pyrenees.

B. Leaves linear or subulate; petals about one-third as long as sepals :—*A. Jacquini*, Koch; flowers in fascicles; hill-sides; Switzerland, Jura, Tirol.

C. Leaves linear or subulate; petals about as long as sepals :—*A. mucronata*, L. (*rostrata*, Koch); cæspitose, flowers in a terminal corymb, sepals white; high; Southern Switzerland, Tirol. *A. stricta*, Whlb. (*uliginosa*, Schleich.); plant glabrous, flowers 3–5, flower-stalk very elongated; peat-swamps; Jura, rare. *A. recurva*, Whlb.;

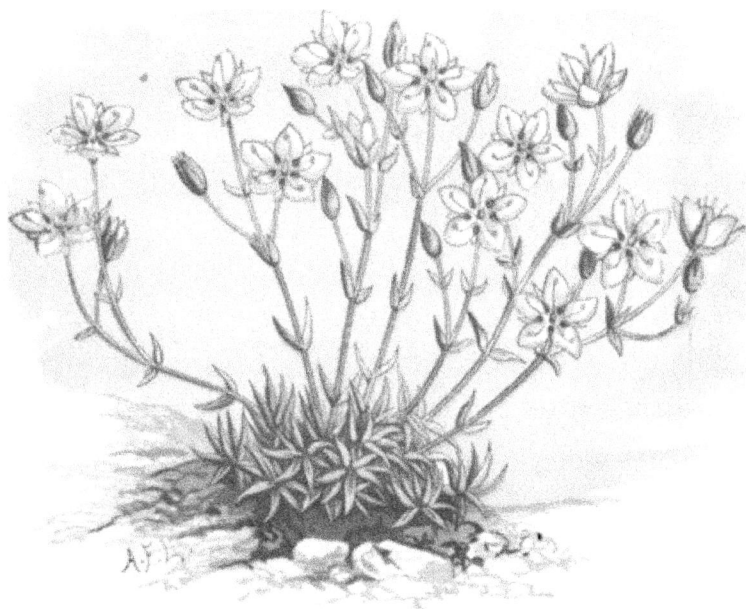

XXI.—ALSINE VERNA.

plant glandular-hairy, flowers 1–3, flower-stalk short, leaves linear, thick, reflexed; high, frequent. *A. verna*, Bartl. (Pl. 21), (*Gerardi*. Whlb., *nivalis*, Fnzl.); cæspitose, plant glandular-hairy, flowers more numerous, leaves flat; Jura, Pyrenees.

D. Leaves linear or subulate; petals 1½ to twice as long as sepals:—*A. laricifolia*, Crntz. (*striata*, Gren.); flowers large (⅓–½ in.), sepals linear, oblong, obtuse, calyx not glandular, capsule as long as calyx; Alps, Jura, Dauphiny, Pyrenees. *A. liniflora*, Heg.; flowers large, sepals linear-oblong, obtuse, calyx glandular, capsule longer than sepals; Jura. *A. biflora*, Whlb.; sepals linear-lanceolate, acute, flower-stalk 1–2-flowered, flower about ⅓ in.; Southern Switzerland, Tirol, very rare. *A. austriaca*, M.K.; sepals linear-lanceolate, acute, flower-stalk very long, 2-flowered; Tirol, Styria, Carniola, Carinthia. *A. Villarsii*, M.K.; sepals linear-lanceolate, acute, flower-stalk very long, 3–7-flowered; Southern Switzerland, Carinthia, Dauphiny, Pyrenees, rare.

Alsine tenuifolia, Crntz., a very slender plant, with petals half as long as sepals, is generally distributed through the lowlands; and *A. viscosa*, Schreb., covered with a glandular pubescence, is occasional in sandy places.

17. CHERLERIA, L.

Petals 5, very narrow or 0; stamens 10; sepals united at the base.

C. sedoides, L. (*Alsine Cherleri*, Fnzl.); densely cæspitose, leaves linear, trigonous, flowers small, sessile, petals green or often wanting; very high; Alps, Pyrenees, Carpathians, frequent.

18. Buffonia, L.

Sepals 4, membranous; petals 4; stamens 4; styles 2; capsule opening with two valves. Plants of a rush-like habit; not alpine.

B. macrosperma, Gay (*paniculata*, Del.); inflorescence compound, crowded, leaves setaceous; gravelly places; Southern Switzerland, Dauphiny.

19. Sagina, L.

Flowers usually small, solitary, on long stalks; sepals 4–5; petals 4–5, often wanting; stamens 4–10; styles 4–5; capsule opening by 4–5 valves; leaves subulate, connate at the base. Small cæspitose inconspicuous plants.

A. Sepals and petals usually 4 :—The English species of Pearlwort, *S. procumbens*, L., with very small petals, and *apetala*, L., usually apetalous, are universally diffused. *S. bryoides*, Rchb., with ciliate leaves, is probably a mountain variety of *procumbens* (Tirol, Carinthia), and *S. ciliata*, Fr. (Southern Switzerland, rare), also with ciliate leaves, a variety of *apetala*.

B. Sepals and petals usually 5 :—*S. repens*, Burn. (*glabra*, Koch); glandular-hairy, petals $1\frac{1}{2}$–2 times as long as sepals; local; St. Bernard, Southern Tirol, Dauphiny. *S. Linnæi*, Presl. (*Spergula saginoides*, L.); petals shorter than sepals, capsule as long as sepals; pastures. *S. macrocarpa*, Maly; petals about as long as sepals, capsule twice as long as sepals; pastures, local. The English species, *S. subulata*, Presl, with the leaves narrowed into a long awn; and *S. nodosa*, Fnzl., with much larger flowers and very swollen stem-nodes, occur also, the former in dry, the latter in moist pastures.

20. SPERGULA, L.

Petals 5; sepals 5; stamens 5–10; styles 5; capsule opening by 5 entire valves; leaves very narrow, with small scabrous stipules. Small herbs; not alpine.

S. arvensis, L., Spurrey; very common in cultivated land.

21. SPERGULARIA, L.

Sepals 5; petals 5; stamens 2–10; styles 3; capsule opening with 3 valves; leaves linear, with scarious stipules. Prostrate herbs; not alpine.

S. rubra, Presl (*Arenaria rubra*, L.), with pink flowers and flat leaves; and *S. media*, Pers. (*marina*, Leb.), with pale pink flowers and fleshy leaves, both with decumbent habit; sandy places. *S. segetalis*, Fnzl., with white flowers; fields; Switzerland, occasional.

Order XIII.—PORTULACACEÆ.

Petals 4 or more; sepals 2; stamens 3 or more; style simple or trifid. A small order, chiefly American; not alpine.

1. MONTIA, L.

Flowers minute, white; petals 5, united at the base; stamens 3; seed-vessel a globose capsule with 1–3 tuberculated seeds.

M. fontana, L.; marshy places; the form *M. minor*,

Gmel., with yellowish-green leaves and opaque seeds, common; *rivularis*, Gmel., with dark green leaves and shining seeds, in Aargau, Vosges, and Black Forest.

Order XIV.—ELATINACEÆ.

Flowers small, in the axils of the leaves; sepals and petals 2–5; stamens 2–10; styles 2–5; seed-vessel a septicidal many-seeded capsule.   A very small order; not alpine.

1. ELATINE, L.

Flowers minute; sepals and petals 2–4.   Very inconspicuous creeping plants, growing in very wet places or under water.

E. Alsinastrum, L.; leaves in whorls; Western Switzerland. *E. Hydropiper*, L.; leaves opposite, flowers sessile, sepals and petals 4 each; Western Switzerland. *E. hexandra*, DC.; leaves opposite, flowers stalked, sepals and petals 3 each, stamens 6; Lake of Geneva. *E. triandra*, Schk.; resembling the last, but stamens 3; Jura.

Order XV.—LINACEÆ.

Inflorescence cymose, flowers regular; sepals 4–5, sometimes united at the base; petals 4–5, distinct; stamens 4–10; seed-vessel a septicidal capsule; leaves always simple and entire.   A small order, widely diffused.

1. LINUM, L.

Flowers in dichotomous cymes, very fugacious; sepals, petals, and stamens 5 each; capsule splitting into 5

valves containing 10 oily seeds. The species of Flax are mostly lowland plants.

A. Flowers small, white; leaves opposite:—*L. catharticum*, L.; dry banks, very common.

B. Flowers large, bright blue:—*L. viscosum*, L.; stigmas club-shaped, sepals glandular-ciliate, leaves pubescent; Tirol, Carinthia, Eastern Pyrenees. *L. narbonense*, L.; stigmas filiform, sepals lanceolate-acuminate, flowers very large (1 in. diam.); Tirol, Styria, Carniola, Pyrenees. *L. alpinum*, L. (Pl. 22); stigmas capitate; outer sepals lanceolate-acuminate, inner ovate, obtuse, stems several, wiry, prostrate or ascending; alpine pastures, local; Jura, Dauphiny, Pyrenees. *L. austriacum*, resembling the last, but with deflexed flower-stalks; Eastern Pyrenees. *L. perenne*, L.; sepals ovate, with membranous border, stems 1–2; Carniola, Pyrenees.

C. Flowers lilac or pale pink:—*L. tenuifolium*, L.; flowers pale lilac, stigmas capitate, sepals glandular-ciliate, sepals elliptical, subulate; local; Switzerland, Pyrenees. *L. suffruticosum*, L.; flowers pale pink, stigmas capitate, stem tortuous, woody, pubescent; Pyrenees.

D. Flowers yellow:—*L. flavum*, L.; Eastern Pyrenees, Lombardy.

2. RADIOLA, Gml.

Flowers very small, in corymbose cymes; sepals, petals, and stamens 4 each; ovary 4-celled; styles 4; leaves opposite. Not alpine.

R. linoides, Gmel. (*Millegrana*, Sm.), a very inconspicuous plant with minute white flowers; wet sandy places, occasional.

Order XVI.—MALVACEÆ.

Flowers regular; calyx 5-lobed, often with an epicalyx; petals 5, twisted in the bud; stamens numerous, the filaments united into a tube surrounding the numerous styles; seed-vessel composed of many carpels; leaves stipulate. A large order, chiefly tropical and subtropical; no alpine species.

1. MALVA, L.

Flowers axillary, conspicuous, fugacious, subtended by an epicalyx of 3 bracts; seed-vessel a flat whorl of numerous 1-seeded indehiscent carpels.

The two common English Mallows, *M. sylvestris*, L., and *rotundifolia*, L., are widely distributed; and *M. moschata*, L., Musk Mallow, with lighter coloured flowers and more divided leaves, occurs occasionally. Also *M. Alcea*, L.; bracts of epicalyx ovate, acute, leaves deeply divided, plant hairy; road-sides, frequent.

2. ALTHÆA, L.

Flowers axillary or in racemes, large; epicalyx of 6–9 bracts.

A. hirsuta, L.; flowers 1 in. diam., pink, on long stalks, leaves kidney-shaped, on long stalks, plant hispid; cultivated land, occasional. *A. officinalis*, L., Marsh Mallow; flowers 1–2 in. diam., pink, in axillary cymes, leaves on short stalks, a pubescent shrub; marshes, but usually introduced.

Other species of *Althæa*, also of *Lavatera*, L., some of

them shrubby, are found in the Departments of the Pyrenees, but are Mediterranean plants.

Order XVII.—TILIACEÆ.

Trees with alternate stipulate leaves; flowers in cymes, nectariferous; sepals and petals usually 5, stamens numerous; ovary free, 2–10 celled. A moderately large order, chiefly tropical; not alpine.

1. TILIA, L.

Flowers in axillary or terminal cymes; peduncle with a leafy decurrent bract; leaves oblique, cordate; seed-vessel globose, indehiscent, 1–2 seeded.

The two English wild species of Lime or Linden, *T. platyphyllos*, Scop., with larger leaves, pubescent beneath; and *parvifolia*, Sm. (*ulmifolia*, Scop.), with smaller glabrous leaves, occur throughout Central Europe; also *T. vulgaris*, Hayne, our cultivated species, with leaves generally glabrous, but pubescent in the axils of the veins beneath.

Order XVIII.—ACERACEÆ.

Trees with opposite exstipulate leaves; flowers in racemes or corymbs, often imperfect; calyx 5-fid; petals 5 or 0; stamens 8 or more; fruit of two or more spreading samaras. A small order, widely distributed; not alpine.

1. ACER, L.

Leaves simple, entire or lobed.

A. campestre, L., Maple, occurs generally in woods and

hedges; and *A. pseudo-platanus*, L., Sycamore, in woods. Also the following:—*A. opulifolium*, Vill. ; corymbs at last drooping, wings of samara nearly parallel, leaves 5-lobed ; mountain woods ; Jura, Vosges, Dauphiny, Pyrenees. *A. platanoides*, L. ; inflorescence corymbose, leaves very thin, with acuminate lobes ; mountain woods ; Jura, Vosges, Dauphiny, Pyrenees. *A. monspessulanum*, L. ; leaves thick, 3-lobed, lobes nearly entire, flowers appearing before the leaves ; Pyrenees, Geneva.

Order XIX.—GERANIACEÆ.

Flowers regular; sepals 5; petals 5; stamens 5 or 10; seed-vessel a beaked capsule, composed of 1-seeded carpels which separate by their base from the beak. A large order, generally distributed, with but few alpine species.

1. GERANIUM, L.

Flowers on axillary flower-stalks; stamens 10; leaves stipulate.

The following British species of Crane's Bill occur in similar situations in Central Europe:—The two large meadow species, *G. pratense*, L., with purple, and *sylvaticum*, L. (sub-alpine), with reddish-purple flowers ; *G. sanguineum*, L., with large crimson flowers and nearly orbicular 5–7-partite leaves ; open pastures, especially calcareous, local. *G. Robertianum*, L., Herb-Robert, with fœtid smell ; everywhere. *G. pyrenaicum*, L. (*perenne*, Huds.), with smaller violet-red flowers and kidney-shaped leaves ; frequent. *G. rotundifolium*, L., a hairy plant, with soft nearly orbicular 7–9-lobed leaves ; stony places,

local. *G. columbinum*, L., with smooth leaves, and flowers in pairs on very long and slender flower-stalks. *G. dissectum*, L., with deeply-divided smooth leaves, and flowers on much shorter flower-stalks. *G. lucidum*, L., leaves orbicular, 5-lobed, glabrous, very shining, tinged with red, flowers small, bright red; rocky places, rare. *G. molle*, L., and *pusillum*, L., small very pubescent plants, with small flowers, the former bright pink, the latter paler; common in cultivated and waste land.

The following are nearly all not British, and are mostly alpine :—

A. Flowers white with red veins :—*G. rivulare*, Vill. (*aconitifolium*, L'Hér.); leaves mostly opposite, with narrow rhomboidal segments and membranous acuminate stipules; rare; Grisons, Valais, Tirol, Dauphiny.

B. Flowers red, pink, or purple; leaves polygonal in their general outline, with rhomboidal lobes :—*G. bohemicum*, L.; petals violet-blue, broadly wedge-shaped, emarginate, ciliate, plant covered with long spreading glandular hairs; very local; Valais, Tirol, Pyrenees. *G. nodosum*, L.; petals light pink, wedge-shaped, deeply emarginate, ciliate, sepals with a long awn, plant finely pubescent; Southern Switzerland, Dauphiny, Pyrenees, rare. *G. phæum*, L.; petals dark violet, nearly entire, ciliate on the claw, plant covered with soft hairs; mountain woods; Western Switzerland, Dauphiny, Pyrenees, rare. *G. palustre*, L.; petals large, purple, entire, ciliate on the claw, fruit-stalk reflexed, plant covered with non-glandular hairs; wet places; Switzerland, Pyrenees. *G. purpureum*, Vill.; flowers small, petals glabrous, resembling *Robertianum*, but with smaller flowers and not fœtid; hedge-banks; Switzerland, Dauphiny.

C. Flowers red, pink, or purple; leaves nearly orbicular in their general outline, 5–7-lobed:—*G. cinereum*
Cav. (Pl. 23); flowers large, pale pink veined with purple,
plant 2–6 in. high, almost stemless, leaves very deeply
divided, grey-green; Central Pyrenees, rare. *G. argenteum*, L.; flowers large, pale pink veined with purple,
petals slightly emarginate, plant grey-green, covered with
silky hairs; high; Tirol, Carniola, Dauphiny, Pyrenees,
rare. *G. divaricatum*, L.; flowers small, pink, leaves
pale green, unequilateral, with one of the lateral lobes
larger than the other, plant 1½–2 in. high, pubescent,
viscous; road-sides; Southern Switzerland, Pyrenees.
G. macrorhizon, L.; flowers reddish-purple, petals narrowed at the base, stem 1–2 ft., springing from an oblique
or horizontal root-stock; Tirol, Styria, Carinthia, Carniola, local.

2. ERODIUM, L'Hér.

Resembling *Geranium;* but stamens 5; flowers usually
in umbels.

Two English species of Stork's Bill, *E. cicutarium*, L.,
with deeply cut hairy leaves, and *moschatum*, L., with soft
pinnate leaves, the whole plant pubescent and smelling of
musk, occur in sandy places, the latter local.

The following sub-alpine species are found in the
Pyrenees :—

E. Manescavi, Boub.; flowers large, ¾–1 in. diam.,
bright crimson, forming an umbel with a large bract
at the base, petals emarginate, sepals acuminate, leaves
all radical, pinnate, with large linear-lanceolate stipules,
flowering stems springing from the root, 12 to 15 in.
high, whole plant hairy. *E. petræum*, Willd. (Pl. 24);

XXIV.—ERODIUM PETRÆUM.

flowers large, pink, in small umbels with two small bracts at the base, flowering stems springing from the root, leaves all radical, bipinnate, a hairy plant with a goaty smell. *E. macradenum*, L'Hér., resembling the last, but smaller, less hairy, with a very penetrating odour.

Order XX.—OXALIDEÆ.

Flowers regular; seed-vessel a capsule, not beaked, splitting by elastic valves; otherwise like Geraniaceæ. A very small order; not alpine.

1. OXALIS, L.

Flower-stalks axillary, one- or several-flowered; leaves trifoliolate; whole plant acid.

Our English Wood Sorrel, *O. Acetosella*, L., is found everywhere in woods. *O. stricta*, L., with yellow flowers in small umbels, exstipulate leaves, and underground stolons; and *O. corniculata*, L., with yellow flowers in small umbels, leaves with small stipules, and no stolons, are weeds in cultivated land.

Order XXI.—BALSAMINEÆ.

Flowers irregular; sepals coloured, the outer one spurred; seed-vessel a capsule. A very small order, chiefly Indian; not alpine.

1. IMPATIENS, L.

Sepals 3 or 5, two of them very small; petals 3; anthers coherent; capsule splitting elastically, and throwing out the seeds to a distance, the valves then twisting.

The beautiful Touch-me-not, *I. noli tangere*, L., is frequent in damp rocky woods. In addition to the handsome pitcher-shaped pale yellow flowers dotted with red, it bears very small inconspicuous apetalous flowers, with a cap-like calyx, which are self-fertilised, and which produce fertile capsules in the autumn.

Order XXII.—HYPERICACEÆ.

Flowers regular, usually in terminal cymes; sepals and petals 5 each; stamens very numerous, the filaments united in their lower part into 3 or 5 bundles; ovary of 3 or 5 carpels, with the same number of styles; seed-vessel usually a capsule; leaves opposite, often dotted. A moderate-sized order, belonging chiefly to the temperate zone; very few alpine species.

1. HYPERICUM, L.

Petals usually oblique, always yellow; leaves sessile, often dotted with glands.

A. Stamens in three bundles; leaves in whorls of 3 or 4:—*H. Coris*, L.; petals glandular-ciliate, leaves linear; rocky slopes, rare; Glarus, Tirol.

B. Stamens in three bundles; leaves opposite. To this section belong nearly all the British species of St. John's Wort, most of which are also found in Switzerland, viz.:—*H. perforatum*, L., with 2-ridged stem and very glandular leaves; very common. *H. quadrangulum*, L., with square stem and bright yellow petals spotted with black. *H. tetrapterum*, Fr., with square stem and pale yellow petals, not spotted. *H. pulchrum*, L. (Northern

Switzerland); glabrous, the buds orange-crimson. *H. hirsutum*, L.; pubescent, flowers small, pale yellow. *H. montanum*, L.; nearly glabrous, flowers pale yellow, the upper pairs of leaves very distant. *H. humifusum*, L.; stem slender, prostrate; very local. *H. elodes*, Huds. (*Elodes palustris*, Sp.); leaves orbicular, woolly, flowers pale yellow, campanulate, half-closed; bogs, very rare. The following are not British:—*H. Richeri*, Vill. (*Burseri*, Sp.), (Pl. 25); leaves ovate-lanceolate, not dotted, flowers few, large, petals spotted, sepals with a broad fringe of glands, terminating in a club, stem 2-ridged; rare; Jura, Dauphiny, Pyrenees. *H. nummularium*, L. (Pl. 26); flowers solitary or very few, leaves coriaceous, glabrous, orbicular, not dotted, stem glabrous, 4–12 in., whole plant tinged with red; high; Dauphiny, Pyrenees.

C. Stamens in five bundles; leaves opposite:—*H. Androsæmum*, L. (*Androsæmum officinale*, All.); shrubby, flowers large, fruit a black berry; woods; Southern Switzerland, Pyrenees.

Order XXIII.—RUTACEÆ.

Flowers usually regular; sepals and petals usually 4 each; stamens usually 10; leaves with translucent dots; seed-vessel generally composed of a number of follicles. A large order, chiefly of tropical and sub-tropical shrubs and trees; not alpine.

1. RUTA, L.

Flowers regular; leaves deeply divided; shrubs with a strong odour.

R. graveolens, L. (including *divaricata*, Ten.), Rue; flowers yellow, leaves somewhat fleshy, glaucous, divided into obtuse oval lobes; a shrub 2–3 ft. high; rocky places; Neuchâtel, Styria, Carniola, Pyrenees. *R. montana*, L.; leaves divided into linear lobes; a small shrub with smaller yellow flowers; Pyrenees.

2. DICTAMNUS, L.

Flowers irregular; sepals and petals 5 each; stamens 10, inclined towards the base.

D. albus, L.; flowers pink, leaves unequally pinnate; stony hills; Southern Switzerland, Dauphiny, Pyrenees.

Order XXIV.—ILICINEÆ.

Flowers small, in axillary cymes; sepals 3–6; petals and stamens 4–5; ovary 3- or more-celled; fruit a drupe with 3 or more 1-seeded stones. A small order of usually evergreen trees and shrubs, widely distributed; not alpine.

1. ILEX, L.

Calyx 4–5-parted, persistent; stamens 4; drupe 4-seeded.

I. Aquifolium, L., Holly; common.

CLASS II.—CALYCIFLORÆ.

Flowers usually with both calyx and corolla; petals (when present) distinct; stamens springing from the calyx or upper part of the ovary (Orders XXV.-XL.).

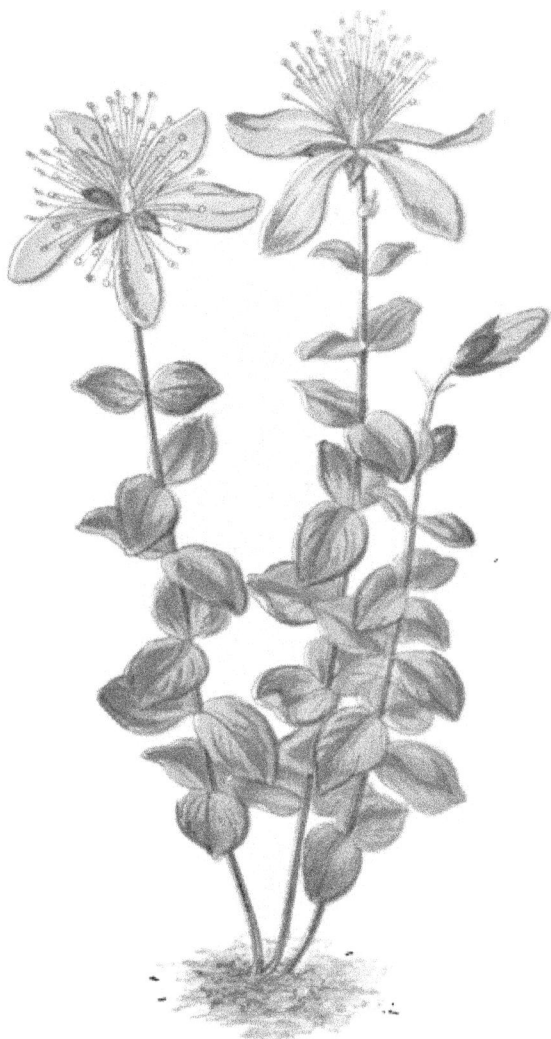

XXVI.—HYPERICUM NUMMULARIUM.

Order XXV.—CELASTRACEÆ.

Flowers small, in cymes; calyx 4–5-lobed; petals and
stamens 4–6; ovary 3–5-celled; seeds usually with an
aril. A small order of trees or shrubs of temperate and
tropical climates; not alpine.

1. Euonymus, L.

Leaves opposite, simple, persistent, with fugacious
stipules; seeds with a conspicuous red aril.

E. europæus, L., Spindle-tree, with square branches
and usually 4 petals; woods and hedges, common.
E. latifolius, Scop., with cylindrical branches, usually
5 petals, and larger leaves; at a somewhat higher
elevation.

Staphylea pinnata, L., Bladder-nut, is hardly wild in
Switzerland.

Order XXVI.—RHAMNACEÆ.

Flowers small, green or yellow; calyx-teeth 4–5, valvate
in bud; petals 4–5 or 0; stamens 4–5; leaves simple,
stipulate. A rather large order of trees and shrubs belonging to warm and temperate regions; very few alpine
species.

1. Rhamnus, L.

Flowers in small axillary cymes, often imperfect; fruit
a drupe enclosed in the calyx-tube.

A. Some or all of the leaves opposite; branches ending
in a spine:—*R. catharticus*, L., Buckthorn; leaves ovate,

serrate, leaf-stalk 2–3 times as long as the deciduous
stipules; stony woods, frequent. *R. saxatilis*, L.; leaves
with short stalks, elliptical, crenulate; stony hills, rare;
Zürich, Dauphiny.

B. Leaves all alternate; branches not spiny: — *R.
Frangula*, L. (*Frangula Alnus*, Mill.), Alder Buckthorn;
leaves elliptic, entire, branches spotted with white;
thickets; grown for the manufacture of gunpowder. *R.
alpinus*, L.; calyx-teeth triangular, leaves ovate, acuminate, crenulate; a small erect shrub; rocky alpine situations. *R. carniolicus*, Kern.; resembling the last, but
leaves narrowly lanceolate; Styria, Carinthia. *R. pumi-
lus*, Turr. (*rupestris*, Scop.); calyx-teeth lanceolate,
acuminate, leaves lanceolate, nearly entire; a prostrate
shrub; rocky alpine situations, local.

Order XXVII.—TEREBINTHACEÆ.

Flowers regular; calyx-teeth 3–5, imbricate in bud;
petals and stamens 3–5; fruit a drupe or berry. A small
order of chiefly tropical trees and shrubs, with a resinous
balsamic juice; not alpine.

1. RHUS, L.

Calyx-teeth, petals, and stamens 5 each; styles 3;
fruit a drupe containing a single seed.

R. Cotinus, L., Sumach; flowers in a loose terminal
panicle, petals yellowish, drupe brown when ripe, leaves
alternate, simple, entire; hills; Southern Switzerland,
Dauphiny, Pyrenees.

Order XXVIII.—LEGUMINOSÆ.

Sepals 5, connate; seed-vessel a legume.   All the
European genera (with two exceptions) belong to the
sub-order PAPILIONACEÆ.   Flowers irregular, pea-like
(standard, keel, and two wings); stamens 10, all united
into a sheath, or the uppermost distinct; leaves alternate,
stipulate (stipules sometimes deciduous), nearly always
compound.   One of the largest natural orders, distributed over every part of the globe, in all climates and
at all altitudes.

Tribe GENISTEÆ.—Leaves simple or of three leaflets,
without tendrils, sometimes 0; all the 10 filaments united
(monadelphous); mostly shrubs.   Genera 1–5.

1. ULEX, L.

Young leaves of 3 leaflets, later ones transformed into
spines; flowers yellow, axillary; calyx membranous, bipartite.   Spiny shrubs; not alpine.

U. europæus, L., Gorse, Furze; Geneva, Ticino; very
rare.

2. SPARTIUM, L.

Calyx resembling a spathe, ending in 5 small teeth;
standard orbicular.   Not alpine.

S. junceum, L.; flowers large, yellow, appearing before
the leaves, leaves few, scattered, with oblong or linear
leaflets; a shrub 3–5 ft., with slender branches, not spiny;
Pyrenees.

3. SAROTHAMNUS, Wimm.

Calyx 2-lipped, 5-toothed, lips divergent; standard orbicular; stigma capitate.   Not alpine.

S. scoparius, Koch (*vulgaris*, Wimm., *Cytisus scoparius*, Lk.), Broom; flowers solitary or in pairs, large, yellow, lower leaves of 3 leaflets, upper simple; a shrub with slender branches; Southern Switzerland, Dauphiny, Pyrenees; not common.

4. GENISTA, L.

Calyx 2-lipped, 5-toothed, lips connivent; flowers yellow, in bracteate racemes; leaves simple; stipules minute or o.   Small shrubs.

A. Stem spiny:— *G. germanica*, L.; leaves ciliate, standard pubescent, upper lip of calyx divided almost to its base; wood-sides in the lowlands. *G. Scorpius*, DC.; stem angular, few-leaved, leaves with two small spiny stipules, standard glabrous; Pyrenees.

B. Not spiny; stem leafless, winged :—*G. sagittalis*, L.; stem prostrate, 2-edged, leaves ovate-lanceolate, flowers in ovate terminal racemes; stony hill-sides, frequent.

C. Stem not spiny, leafy :—*G. pilosa*, L.; stem prostrate, tortuous, leaves silky beneath, standard and keel silky, pod pubescent; heathy places; Jura, Carinthia, Styria, Pyrenees. *G. decumbens*, L.; prostrate, upper calyx-lip with 2 short teeth, flower-stalk about three times as long as calyx; Jura. *G. tinctoria*, L.; stem erect, 1–2 ft., rigid, leaves adpressed, ciliate, flowers large, ½ in, pod glabrous; meadows, widely distributed. *G. mantica*, Poll.; resembling the last, but with hairy pods; Ticino. *G. Delarbrei*, L. and L.; similar, but a larger plant, with larger flowers and pods; Jura. *G. ovata*, W.K.; similar, but stem and leaves hairy, leaves broader, and racemes shorter; Styria. *G. cinerea*, DC.; stem $1\frac{1}{2}$–3 ft. high, leaves silky, the whole plant grey-green, calyx and standard silky; high; Pyrenees, Dauphiny. *G. triangularis*, Willd.; glabrous, branches triquetrous, ascending, leaves lanceolate, mucronate, flowers axillary; Styria.

5. CYTISUS, L.

Leaves usually of three leaflets, with minute stipules; flower usually yellow; calyx-teeth very small; wing-petals deflexed after flowering; style incurved. Shrubs or trees; not generally alpine.

A. Leaves opposite :—*C. radiatus*, Koch; stem 2–3 ft. high, leaflets linear, deciduous; Valais, Southern Tirol, Carinthia, rare.

B. Leaves alternate; raceme drooping; calyx campanulate; trees:—*C. Laburnum*, L., the Laburnum; ovary and pod silky, leaves with adpressed hairs; mountain woods; Jura, Ticino, Pyrenees. *C. alpinus*, Mill.; flowers dark golden yellow, ovary and pod glabrous, leaflets hairy at the margin; mountain woods; Jura, Southern Switzerland, Carniola, Dauphiny.

C. Leaves alternate; raceme not drooping; calyx campanulate; shrubs :—*C. decumbens*, Walp.; stem decumbent, branched, raceme unilateral, calyx hairy, pod black when ripe; Jura.

D. Leaves alternate; raceme not drooping; calyx tubular; shrubs :—*C. hirsutus*, L.; flowers 2–4 in., in lateral fascicles, calyx hairy, standard glabrous, branches leafy, covered with spreading hairs; Southern Switzerland, Dauphiny. *C. capitatus*, Jacq.; flowers in a dense terminal corymb, stem erect, covered with spreading hairs, calyx hairy; Jura, Pyrenees. *C. supinus*, L.; resembling the last, but stem prostrate, flowers only 2-4, smaller; Dauphiny, Pyrenees. *C. glabrescens*, Sart.; glabrous, flowers in lateral fascicles, with long flower-stalks; rare; Ticino, Como. *C. nigricans*, L.; racemes erect, leafless, stem 3-4 ft.; stony slopes; Switzerland. *C. austriacus*, L.; racemes terminal, stem erect, pubescent, leaves lanceolate, calyx and legume slightly hairy; Carniola.

E. Flowers purple:—*C. purpureus*, Scop.; flowers solitary, axillary, on short stalks, stem ascending, nearly simple, calyx, leaves, and legume glabrous; Tirol, Styria, Carniola, Carinthia.

Tribe TRIFOLIEÆ.—Leaves trifoliolate, with usually adnate stipules, without tendrils; uppermost filament usually distinct (diadelphous). Mostly herbaceous. Genera 6-11.

6. ONONIS, L.

Flowers solitary or few; filaments monadelphous. Mostly prostrate hairy herbs or small shrubs.

A. Flowers yellow; sub-alpine:— *O. Natrix*, L.; flowers large, in terminal leafy racemes, standard orbicular, streaked with red, stem-leaves sometimes with 5-7 leaflets, leaflets oblong, finely dentate, stalked, plant glandular-viscid and fœtid; Southern Switzerland, Carniola, Styria, Pyrenees. *O. Columnæ*, All.; flowers solitary in the axils of the leaves, small, pale yellow, corolla not longer than the calyx, stem woody, glandular-pubescent; hills; Southern Switzerland, Pyrenees. *O. minutissima*, L.; flowers smaller, crowded, corolla bright yellow, but the spring flowers often apetalous, autumn flowers in loose spikes; Dauphiny, Pyrenees.

B. Flowers red or purple:—Our two common English species of Rest-harrow, *O. spinosa*, L. (*campestris*, Koch), and *arvensis*, L. (*repens*, Koch, *procurrens*, Wall.), are common in sandy places. The following are sub-alpine: —*O. rotundifolia*, L.; flowers 2–3 in the axils of the leaves, stalked, leaflets all stalked, nearly round, dentate, plant glandular-pubescent; high; Jura, Southern Switzerland, Dauphiny, Pyrenees. *O. fruticosa*, L.; flowers large, purple, in fascicles of 2 or 3, leaves glabrous, leaflets oblong, sessile, strongly toothed; a glabrous shrub 1–2 ft.; high; Dauphiny, Pyrenees. *O. cenisia*, L.; flowers solitary in the axils of the upper leaves, leaflets small, coriaceous, sessile, glabrous, stem slender, prostrate; high; Dauphiny, Pyrenees.

7. TRIGONELLA, L.

Filaments monadelphous; calyx tubular; legume longer than the calyx, curved. Not alpine.

T. monspeliaca, L.; flowers yellow, in few-flowered nearly sessile umbel-like racemes; dry hills; Southern Switzerland, Dauphiny.

8. MEDICACO, L.

Flowers small, yellow or violet, in very short racemes; filaments diadelphous; legume coiled or curved spirally, often spiny. Not alpine.

The two commonest English species of Melick—*M. Lupulina*, L., and *minima*, Desr.—are common in grassy situations. Others have been introduced, including *M. sativa*, L., Lucerne, with larger violet flowers. *M. falcata*, L., a partially erect plant with violet-yellow flowers and sickle-shaped legume, grows by road-sides or in meadows. *M. suffruticosa*, Ram., a pubescent plant, 2–8 in. high, with rather large yellow flowers, round or obovate leaflets, and half-arrow-shaped stipules, is found in rocky sub-alpine situations in Pyrenees.

9. MELILOTUS, Tourn.

Flowers small, yellow or white, in long axillary racemes; filaments diadelphous; legume short, straight, few-seeded. Not alpine.

M. alba, Desr. (*vulgaris*, Willd.), with white flowers, is a very common plant by road-sides. *M. officinalis*, Desr. (*arvensis*, Wallr.), Melilot, and *altissima*, Thuill., with yellow flowers, are also widely distributed.

10. TRIFOLIUM, L.

Flowers usually in round umbel-like spikes or racemes, containing honey; petals united into a long tube; legume small, straight, few-seeded.

Most of the English species of Trefoil are Swiss; there are also some alpine species.

A. Flowers purple or nearly white, changing to pink.

The following are English:—*T. pratense*, L., Red Clover; *T. medium*, L., with larger green stipules; *T.*

fragiferum, L., Strawberry Trefoil, with very light pink flowers, and calyx greatly swollen in fruit; *T. arvense*, L., Hare's-foot Trefoil, very hairy, with small pink flowers; *T. striatum*, L., hairy, procumbent, with minute flowers; *T. rubens*, L., with glabrous stem and leaves, and oblong-cylindrical flower-heads; *T. hybridum*, L., with flowers at first white, changing to pink, and erect hollow stem, in damp meadows; and *T. elegans*, Sav., with smaller heads than the last, and flowers pink from the first, are lowland species.

The following are alpine or sub-alpine :—*T. alpinum*, L.; stemless, flowers large, up to 1 in. in length, light red, in 3–8-flowered spikes, lower tooth of calyx longer than the others, leaflets linear, serrate; alpine pastures, frequent. *T. alpestre*, L.; stem short, flowers smaller, spike many-flowered, calyx-tube hairy, leaflets denticulate; alpine pastures; Switzerland, Pyrenees. *T. Thalii*, Vill.; stemless, flowers first white then pink, in many-flowered spikes, calyx-teeth nearly equal; pastures, especially Southern, local. *T. montanum*, L.; flowers very pale pink, flower-stalk deflexed after flowering, calyx-teeth equal, leaflets hairy beneath; alpine pastures.

B. Flowers white or yellowish white :—*T. repens*, L., White or Dutch Clover; and *scabrum*, L., a small prostrate plant, with minute white flowers; are English and lowland. The remainder are alpine or sub-alpine :—*T. pallescens*, Schreb.; stem prostrate, flowers yellowish-white, keel finally pink, flower-stalk reflexed after flowering; pastures; Switzerland, Carniola, Tirol, Pyrenees. *T. glareosum*, Schleich.; flowers white, pendant, in axillary long-stalked spikes, leaf-stalk short; pastures, local. *T. ochroleucum*, L.; flowers yellowish-white, calyx-teeth lan-

ceolate-subulate, unequal, the lower one finally bent back ;
pastures ; Switzerland, Pyrenees, local.   *T. saxatile*, All.;
flowers white, in dense globular spikes, surrounded by an
involucre of bracts, calyx very hairy, stipules dotted with
black ; very high, near the glaciers, rare; Simplon, Nicolaithal, Saas, Tirol.   *T. nivale*, Sieb.; flowers white, calyx,
stem, and leaves nearly glabrous, stipules prolonged into a
sharp awn ; Switzerland, very high, local.   *T. noricum*,
Wulf. ; flowers white, stem and leaves woolly, stipules
acuminate ; Southern Tirol, Styria, Carniola, Carinthia.
T. pannonicum, Jacq.; flowers yellowish-white, upper
leaves opposite, plant very hairy; Piedmont, Carniola.

   C. Flowers yellow :—*T. procumbens*, L. (*campestre*,
Schreb.), Hop Trefoil, with dense heads of pale yellow
flowers ; and *T. minus*, Sm. (*dubium*, Sibth.), with much
smaller heads of smaller flowers ; both prostrate plants ;
are common lowland English species.   To the same section belong *T. patens*, Schreb. ; heads composed of 5–15
stalked flowers, standard streaked, stipules auriculate;
Ticino, Pyrenees, Dauphiny ; *T. agrarium*, L.; heads composed of 20–50 nearly sessile flowers, standard strongly
streaked, stipules rounded at the base; Pyrenees; and
T. aureum, Poll., similar, but stipules not rounded at the
base ; Jura, Dauphiny, Pyrenees.

   The following are sub-alpine :—*T. spadiceum*, L.; upper
leaves opposite, heads finally elongated, cylindrical, flowers
turning deep chestnut brown ; high pastures ; Switzerland, Dauphiny, Pyrenees.   *T. badium*, Schreb. ; similar,
but heads finally nearly globular, flowers turning light
brown ; Jura, Carpathians, Dauphiny, Pyrenees.

11. DORYCNIUM, Tourn.

Leaves sessile; stipules free, resembling the leaflets; pod swollen, 2–6 seeded.

D. suffruticosum, Vill.; stem shrubby, leaves linear-wedge-shaped, flowers small, white, keel dark blue at the tip; rare; Chur, Southern Tirol, Pyrenees. *D. herbaceum*, Vill.; stem herbaceous, leaves oblong, wedge-shaped, keel blue throughout; Ticino, Dauphiny.

Tribe LOTEÆ.—Leaves of 5 or more leaflets, one terminal; stipules small or o; alternate filaments dilated; legume without a longitudinal septum. Genera 12–14.

12. ANTHYLLIS, L.

Flowers in capitate cymes, sometimes with an involucre; calyx inflated; petals with long claws, standard auricled at the base; legume included in the calyx.

A. Vulneraria, L., Kidney Vetch, Lady's Fingers; flowers yellow or orange, pairs of leaflets 1–5, terminal leaflet much the largest, whole plant very silky; pastures, common. *A. montana*, L.; flowers red, pairs of leaflets 10–20, all nearly equal in size, plant covered with silky wool; rocky places; Jura, Tirol, Styria, Pyrenees.

13. LOTUS, L.

Leaflets usually 5; calyx not inflated; flowers yellow, uppermost filament distinct; legume septate between the seeds. Not alpine.

L. corniculatus, L.; leaflets obliquely obovate, very

common. *L. tenuis*, L.; leaflets oblong-linear, flowers
few; damp places; *L. uliginosus*, Schk. (*major*, Sm.); a
larger plant, with the racemes on very long stalks; bushy
places; are all English and lowland.

14. TETRAGONOLOBUS, Scop.

Leaves of 3 leaflets; stipules free, leaf-like; legume
with 4 membranous wings, coiling up like a corkscrew
after bursting.

T. siliquosus, Roth; flowers large, solitary, sulphur-
yellow, on long stalks; damp meadows; Southern
Switzerland, Pyrenees.

Tribe GALEGEÆ.—Leaflets 5 or more, uppermost
filament distinct; legume 2-valved, with a longitudinal
septum.   Genera 15–18.

15. ASTRAGALUS, L.

Leaves pinnate with a terminal leaflet, leaflets entire;
flowers in axillary often globular spikes or racemes;
bracts small; calyx tubular, teeth small, nearly equal;
petals usually narrow, with long claw, keel obtuse;
style beardless.   An alpine genus.

A. Flowers purple, blue, red, or violet:—*A. monspessulanus*, L. (Pl. 27); flowers red, in spikes of 15–20,
with peduncles at least as long as the leaves, standard
very long and narrow, leaves of 15–20 pairs of leaflets,
stipules acuminate, united to the petiole by half their
length, stem very short, plant 4–8 in.; gravelly places;
Southern Switzerland, Dauphiny, Pyrenees. *A. leontinus,*

XXVII. –ASTRAGALUS MONSPESSULANUS.

Wulf.; flowers light blue or lilac, standard longer than the wings, pairs of leaflets 6–10, stipules coherent, but free from the leaf-stalk, calyx covered with black hairs, plant pubescent, 2–4 in.; Switzerland, Carniola, Tirol, Pyrenees. *A. Onobrychis*, L.; pairs of leaflets 8–12, flowers large, in dense racemes, standard 2–3 times as long as wings, linear, stipules as in the last; high; Southern Switzerland, Dauphiny, Tirol, Carniola. *A. purpureus*, Lam.; flowers blue-purple, in a globular spike, stalk longer than the leaves, calyx campanulate, with spreading hairs, pairs of leaflets 10–12, stipules as in the last; a hairy plant, 4–6 in.; stony; Southern Tirol, Carniola, Pyrenees. *A. hypoglottis* (*danicus*, Retz); flowers purple-blue, calyx with black hairs, leaflets oblong or linear-oblong, stipules connate, stalk of spike very long, plant hairy; a lowland species. *A. vesicarius*, L.; flowers large, violet, spike globular, calyx hairy, pod almost entirely enclosed in the calyx, leaflets small, in 4–7 pairs, plant silver-white; Dauphiny. *A. austriacus*, L.; flowers very small, pale blue spotted with violet, spike loose, oblong, stipules free, plant glabrous; Dauphiny, Carniola. *A. incanus*, L.; flowers light purple, spike globular, stipules linear, partially coherent to the leaf-stalk, plant prostrate, covered with silky hairs; Dauphiny. *A. oroboides*, Horn.; flowers violet, keel shorter than the standard, legume hairy; Styria, Tirol, rare.

B. Flowers white, or white and violet :—*A. australis*, Lam. (*Phaca australis*, L.), wing-petals deeply bifid, leaflets in 4–8 pairs; high pastures. *A. alpinus*, L. (*Phaca astragalina*, DC.), pairs of leaflets 7–11, stipules free, flowers horizontal or pendant, very pale, fragrant, keel about as long as standard, decumbent; high pastures. *A. hamosus*, L.; flowers few, very small, nearly white, in a globular spike, stipules connate, plant pubescent; Dauphiny, Pyrenees.

C. Flowers yellow:—*A. glycyphyllos*, L.; flowers pale yellow, in an elongated raceme, stalk much shorter than the leaves, pairs of leaflets 5–6, stipules very large, stem scrambling, 2–3 ft.; a lowland species. *A. exscapus*, L.; stemless, flowers large, crowded, yellow, stipules adnate to the leaf-stalk; Southern Switzerland, Tirol, local. *A. depressus*, L.; stemless, flowers yellowish-white with violet spots, stipules free, very broad, oval-acuminate; Southern Switzerland, Southern Tirol, Dauphiny, Pyrenees. *A. Cicer*, L.; flowers pale yellow, pairs of leaflets 8–12, legume inflated, nearly globular, plant decumbent, nearly glabrous; lowlands, sandy places; Switzerland, Dauphiny, Pyrenees. *A. aristatus*, L'Hér.; flowers pale yellow, terminal leaflet replaced by a spine, plant prostrate, very leafy; Southern Switzerland, Dauphiny, Pyrenees.

16. PHACA, L.

Leaves pinnate with a terminal leaflet; flowers in axillary racemes; style not bearded; legume somewhat inflated, undivided, with more than 2 valves. Alpine.

P. alpina, Jacq.; flowers yellow, pairs of leaflets 9–12, stipules linear-lanceolate, stem hairy, branched, 12–15 in.; high pastures. *P. frigida*, L.; flowers yellowish-white, pairs of leaflets 4–5, stipules ovate, stem hairy, simple, 8–10 in.; very high; Switzerland.

17. Oxytropis, DC.

Leaves pinnate, with a terminal leaflet; flowers in axillary clusters or racemes; petals with a long claw, keel very acute, erect; style beardless; legume more or less divided longitudinally, many-seeded. Alpine.

A. Flowers yellow:—*O. pilosa,* DC.; legume completely 2-celled, stem 8–15 in., erect, very leafy, pubescent; stipules free; stony, rare; St. Gallen, Dauphiny. *O. campestris,* DC.; legume partially 2-celled, leaflets in about 12 pairs, plant pubescent but not glandular, flowers varying in colour, sometimes a dirty violet (*sordida,* DC.); high, stony, frequent. *O. fœtida,* DC.; leaflets small, usually in about 20 pairs, plant covered with viscid fœtid glands; rare; Valais, Dauphiny.

B. Flowers violet, blue, or pink:—*O. uralensis,* DC. (*Halleri,* Bunge); stemless, legume completely 2-celled, flowers violet, plant downy with silky hairs or woolly (*velutina,* Sieb.); rocky, rare; Southern Switzerland, Tirol, Carinthia, Styria, Dauphiny, Pyrenees. *O. pyrenaica,* G. and G. (Pl. 28); resembling the last, but legume only partially 2-celled (as in the following species); Pyrenees. *O. montana,* Bge.; young fruit pendant, ripe fruit erect, stalk as long as calyx-tube, pod swollen, plant grey with silky hairs (or nearly glabrous, *Jacquini,* Bge.); rare; Jura, Southern Switzerland, Monte Generoso, Dauphiny. *O. carinthiaca,* Tisch.; resembling the last, but young fruit erect, ripe fruit pendant; Carinthia. *O. neglecta,* Gay (*cyanea,* Gaud.); flowers blue, head 6–12-flowered, calyx-teeth longer, plant silky; rare; Nicolaithal, Tirol, Carinthia, Pyrenees. *O. lapponica,* Gay; resembling the last, but with pendant pods; rare; Saas, Simplon, St.

Bernard, Faulhorn. *O. triflora*, Hoppe; flowers pink, head 2–4-flowered, standard emarginate; Carniola, Carinthia, Salzburg.

18. COLUTEA, L.

Calyx campanulate; keel truncate; legume stalked, bladdery.

C. arborescens, L., Bladder Senna; a shrub with large yellow flowers and very swollen pod; Neuchâtel, Uri, Dauphiny.

Tribe HEDYSAREÆ.—Leaves pinnate, with a terminal leaflet; upper filament distinct; legume indehiscent, jointed.   Genera 19–23.

19. CORONILLA, L.

Flowers in umbels; calyx campanulate, upper teeth nearly united; petals clawed; keel acuminate; legume cylindrical or quadrangular.

A. Flowers variegated, red and white; leaflets in 6–12 pairs; plant 12–24 in. high:—*C. varia*, L.; woods; Switzerland, Dauphiny, Pyrenees.

B. Flowers yellow; a large shrub; leaflets in 2–4 pairs; claw of petals very long and narrow:—*C. emerus*, L.; hill-sides, frequent.

C. Flowers yellow; herbaceous or suffruticose:—*C. montana*, Scop. (*coronata*, L.); umbels 15–30-flowered, stipules coherent, membranous, deciduous, plant 16–24 in.; Jura, Grisons, Pyrenees. *C. vaginalis*, Lam.; leaflets in 5–7 pairs, with white edges, stipules large, deciduous, claw of petals as long as calyx, plant 8–18 in., glabrous; rocky places; Switzerland, Jura, Dauphiny.

XXIX.—CORONILLA MINIMA.

C. minima, L. (Pl. 29); leaflets in 3–4 pairs, stipules very small, persistent, plant prostrate, glabrous, 4–8 in. high; rare; Southern Switzerland, Jura, Pyrenees.

20. ORNITHOPUS, L.

Flowers very small, in long-stalked umbels; leaves pinnate, with a terminal leaflet; legume slender, curved, many-jointed; small prostrate herbs. Not alpine.

O. perpusillus, L., Bird's-foot Trefoil; flowers white, bracts pinnate; sandy fields; very rare in Switzerland.

21. HIPPOCREPIS, L.

Flowers in long-stalked umbels, drooping; petals with a long claw; stipules small or 0; legume curved, with 3–6 horseshoe-shaped 1-seeded joints. Not alpine.

H. comosa, L., Horseshoe Vetch; flowers yellow, leaflets in 4–8 pairs; dry banks, common, especially calcareous.

22. ONOBRYCHIS, L.

Flowers in axillary spikes or racemes; wings short; keel obliquely truncate; leaves pinnate, with a terminal leaflet; legume compressed, not jointed, often spiny.

O. sativa, Lam. (*viciæfolia*, Scop.), Sainfoin; flowers pink, striped with red, raceme long-stalked, leaves in 6–12 pairs, stem prostrate; sandy hills. *O. montana*, DC.; flowers smaller, pink or purple, leaflets shorter and broader; alpine pastures; Alps, Jura, Dauphiny, Pyrenees. *O. supina*, DC. (*arenaria*, DC.); flowers small, white streaked with red, legume pubescent, spiny, stem prostrate; sandy hills; Western Switzerland, Pyrenees.

23. HEDYSARUM, L.

Flowers in axillary racemes; keel oblong, truncate, longer than the wing-petals; leaflets numerous; legume compressed, breaking up into many 1-seeded joints.

H. obscurum, L. (Pl. 30); flowers large, purple, pods drooping, stipules connate, stem erect, 8–25 in.; pastures; Switzerland, Dauphiny, Carpathians, Pyrenees.

Tribe VICIEÆ.—Leaves pinnate, ending in a tendril or point; leaflets sometimes 0; upper filament distinct; herbs, sometimes climbing.   Genera 24–25.

24. VICIA, L.

Leaves usually ending in a tendril, with numerous leaflets and half-arrow-shaped stipules; legume compressed, undivided, many-seeded.   Prostrate or climbing herbs, mostly lowland.

A. Flowers few and very small; leaflets few; calyx not gibbous (*Ervum*, L.); weeds in cultivated ground:—
V. tetrasperma, Mœnch.; glabrous, legume shortly stalked, about 4-seeded.   *V. hirsuta*, Koch, Tare; hairy, legume sessile, 2-seeded.

B. Flowers few, sessile or on very short stalks; legume sessile; calyx not gibbous at the base; mostly annual herbs:—*V. sativa*, L., Field Vetch; flowers large, blue or violet.   *V. angustifolia*, Roth; flowers smaller, legume narrower, glabrous, leaflets linear; Western and Southern Switzerland, Pyrenees.   *V. lathyroides*, L.; flower solitary, very small, violet, tendril often wanting, stipules entire; sandy places; very

XXX.—HEDYSARUM OBSCURUM.

rare in Switzerland. *V. pyrenaica*, Pourr.; flowers large, solitary, pink, tendril short or o, leaflets mucronate, stipules half-arrow-shaped, dentate; Dauphiny, Pyrenees.

C. Flowers few, sessile or on very short stalks; legume stalked; calyx gibbous:—*V. peregrina*, L.; flower solitary, purple, leaf terminating in a branched tendril, leaflets linear, mucronate; near Montreux, Styria, Pyrenees. *V. lutea*, L.; flowers solitary or in pairs, sulphur-yellow, standard glabrous, legume hairy; rare; Southern Switzerland, Jura, Pyrenees. *V. hybrida*, L.; resembling the last, but with hairy standard and truncate or mucronate leaflets; rare, Southern Switzerland. *V. grandiflora*, Scop.; flowers solitary or in pairs, standard obovate, mucronate, leaflets 10–12, obcordate, mucronate, stipules ovate, legume dusky; Styria.

D. Flowers in stalked racemes; legume sessile:—*V. narbonensis*, L.; flowers 1–5, dirty purple, stem simple, lower leaves without tendrils, pod black when ripe; Bâle, Pyrenees.

E. Flowers in stalked racemes; legume stalked:—*V. sepium*, L.; flowers dull purple, 2–5, standard glabrous, leaflets in 4–7 pairs, plant nearly glabrous; very common. *V. pannonica*, Jacq.; resembling the last, but flowers bright purple, standard hairy, plant more hairy; Styria. *V. onobrychioides*, L.; flowers 6–12, large, violet, standard glabrous, stipules half-hastate, stem climbing, glabrous; hill-slopes; Southern Switzerland, Pyrenees. *V. dumetorum*, L.; flowers violet-red then yellowish, leaflets oval, apiculate, stipules crescent-shaped, strongly toothed; mountain woods; Switzerland, Vosges, Dauphiny, Pyrenees. *V. pisiformis*, L.; flowers large, greenish-yellow, 10–15 in long-stalked racemes, leaflets in 4–5 pairs, large, oval, stem 3–5 ft., climbing; bushy places, rare; Switzerland, Pyrenees. *V. oroboides*, Wulf.; flowers large, light yellow, 2–6 in sessile racemes, leaves without tendrils, leaflets in 1–3 pairs, mucronate; bushy places, rare; Southern Tirol, Styria, Carniola, Carinthia. *V. sylvatica*, L., Wood Vetch; flowers large, white with blue veins, 6–18 in a long-stalked raceme, stipules crescent-shaped, toothed, stem 2–4 ft., scrambling, glabrous; thickets in the lowlands, common. *V. Orobus*, DC. (*Orobus sylvaticus*, L.); flowers white spotted with violet, stipules half-hastate, slightly toothed, leaves without tendrils, stem erect, 1–2 ft.; woods; Pyrenees. *V. cassubica*, L.; resembling the last, but with a smaller raceme, and with smaller violet flowers; woods; Pyrenees. *V. villosa*, Roth; plant villous, flowers very small, 2–3, stem 4-angled, leaflets linear-oblong, mucronate, in 8–10 pairs, stipules half-arrow-shaped, legume 2-seeded; Southern Switzerland, Tirol.

F. Flowers in stalked racemes; calyx usually gibbous at the base; legume prolonged into a beak :— *V. Cracca*, L.; flowers blue, 15–20 in dense racemes, leaves long and narrow; hedges in the lowlands, common. *V. Gerardi*, G. and G.; resembling the last, but with a somewhat denser raceme and erect hairy stem; thickets; Southern Switzerland, Tirol, Pyrenees. *V. tenuifolia*, Roth; flowers light blue, elongated, in very long lax racemes, calyx not gibbous; thickets; Switzerland, Pyrenees. *V. varia*, Host.; flowers violet, elongated, horizontal, stipules of two kinds, entire and laciniated, stem ascending, 1–2 ft.; cultivated land.

25. LATHYRUS, L.

Resembling *Vicia*, but with broader petals, the leaves usually with fewer leaflets, generally with tendrils. Not alpine.

A. Leaves reduced to a tendril, stipules leaf-like :— *L. Aphaca*, L.; flower solitary, pale yellow; cornfields; Switzerland (rare), Styria, Dauphiny, Pyrenees.

B. Leaf-stalk broad, leaf-like, without leaflets; no tendrils :—*L. Nissolia*, L.; flower solitary, crimson; among grass; Switzerland (rare), Styria, Pyrenees.

C. Leaves pinnate, with tendrils; style twisted, channelled below :—*L. hirsutus*, L.; flowers violet-blue, stalk of raceme much longer than the leaves, legume rough, seeds pitted; cultivated land; Switzerland, Pyrenees, *L. Cicera*, L.; flowers small, pink, pod channelled on the back, seeds smooth, stipules as long as the leaves; cultivated land; Southern Switzerland, Pyrenees, rare.

D. Leaves pinnate, with tendrils; style twisted, tubular:—*L. sylvestris*, L.; flowers large, greenish streaked with pink, leaflets 2, linear-lanceolate, stem climbing, broadly winged; woods, frequent. *L. heterophyllus*, L.; resembling the last, but flowers pink, leaf-stalk broadly winged, stipules larger; mountain woods; Switzerland, Tirol, Dauphiny, Pyrenees. *L. latifolus*, L., Everlasting Pea; flowers large, bright red, stem climbing, stem and leaf-stalk broadly winged; Pyrenees. *L. cirrhosus*, Ser.; flowers purple, leaflets in 2–3 pairs, leaf-stalk not winged, stem climbing, winged, seeds tubercled; Pyrenees, rare. *L. tuberosus*, L.; flowers purple, fragrant, leaflets 2, leaf-stalk not winged, stem angular, not winged, seeds smooth, root with tubers; fields, frequent.

E. Leaves pinnate, usually without tendrils; style straight, channelled below; flowers in small racemes (*Orobus*, L.):—*L. vernus*, Wim.; flowers 5–7, large, blue, leaflets acuminate, stem erect, 8–26 in., angular, not winged, pod glabrous; woods, frequent. *L. luteus*, Gren.; flowers 4–6, large, yellow, leaves bluish-green beneath; woods; Alps, Jura, Dauphiny, Pyrenees. *L. montanus*, G. and G.; flowers large, yellow, numerous, leaflets elliptic or lanceolate, glaucous beneath, stem angular; mountain woods. *L. macrorhizus*, Wimm. (*Orobus tuberosus*, L.); flowers 2–4, reddish, stem narrowly winged, root tubercled; woods, common. *L. palustris*, L.; flowers 2–3, purple, leaf-stalk not winged, stem climbing, winged; marshes; Switzerland, rare. *L. niger*, Wimm.; flowers 4–8, purple, leaf-stalk not winged, leaflets apiculate, stem angular, not winged, plant turning black on drying; woods; Switzerland, Pyrenees, rare. *L. pratensis*, L.; flowers 3–12, yellow, leaflets 2, leaf ending in a tendril; very common. *L. asphodeloides*, G. and G.; flowers 4–8, pale yellow, leaf-stalk winged, stem 8–16 in., erect, angular, not winged; mountain pastures; Dauphiny. *L. canescens*, G. and G. (*ensifolius*, Gay); flowers 4–10, large, white and blue, leaflets 4–6, linear-lanceolate, leaf-stalk very short; mountain pastures; Neuchâtel, Jura, Dauphiny, Pyrenees.

F. Resembling the last, but flowers solitary, leaves ending in a tendril:—*L. angulatus*, L.; flowers purple, veined, flower-stalk 5–6 times as long as leaf-stalk, leaflets 2, linear, seeds cubical, tuberculated, stem angular, not winged; cultivated land; Southern Switzerland, Pyrenees. *L. sphæricus*, Retz; flowers small, pink, leaflets 2, linear-lanceolate, seeds spherical, stem ascending, angular, 4–8 in.; grassy places; Southern Switzerland, Pyrenees.

Order XXIX.—ROSACEÆ.

Flowers regular; calyx inferior or superior, 5-lobed; petals usually 5, rarely 0, distinct; stamens numerous, rarely few, inserted on the calyx-tube; carpels 1 or more, distinct or more or less united, free or adnate to the calyx-tube; ovules 1 or 2 in each carpel. Herbs, shrubs, or trees, with alternate simple or compound stipulate leaves. A very large order, found in all parts of the globe, but most abundant in the cooler.

Tribe PRUNEÆ.—Calyx inferior, deciduous; carpel 1; fruit a drupe (stone-fruit); trees or shrubs with simple leaves; stipules small, deciduous.

1. PRUNUS, L.

Characters of the Tribe.

P. spinosa, L. (*communis*, Huds.), Sloe, Blackthorn; branches all spiny, flowers small, appearing before the leaves; *P. insititia*, L., Bullace; with broader leaves and larger fruit, and the branches not all spiny; *P. Cerasus*, Wild Cherry; flowers large, in umbels of 2–5, leaves crenate, glabrous; *P. avium*, L.; flowers large, in umbels of 2–5, leaves sessile, pubescent beneath; *P. Padus*, L., Bird-Cherry; flowers smaller, in axillary or terminal racemes; all in hedges and thickets, the last three often planted. Also *P. Mahaleb*, L.; flowers in short erect 3–12-flowered racemes, leaves roundish-ovate, slightly cordate; mountain woods; Southern Switzerland, Tirol,

Pyrenees; rare. *P. brigantiaca*, Vill., flowers in umbels of 2–5, leaves doubly dentate, ciliate, branches not spiny, fruit yellow; mountain woods; Dauphiny.

Tribe SPIRÆEÆ.—Calyx inferior, persistent; carpels 5 or more, each ripening into a many-seeded follicle.

2. SPIRÆA, L.

Flowers in axillary or terminal cymes, white or red; stamens 20 or more; carpels 5 or more, free or connate below.

S. Ulmaria, L., Meadow Sweet; flowers white, in very large cymes; common in wet places. *S. Filipendula*, L., Dropwort; with reddish flowers in smaller cymes, the leaves much smaller and chiefly radical; on sunny hillsides, chiefly cretaceous. *S. salicifolia*, L. (*obovata*, Raf.); a shrub with pink flowers in dense cylindrical cymes, stipules 0; Styria, Carinthia. *S. Aruncus*, L.; a shrub with white flowers in very large feathery cymes, and large pinnatifid leaves; mountain woods; Switzerland, Jura, Pyrenees, frequent. *S. ulmifolia*, Scop.; a shrub with white flowers in corymbose cymes, leaves simple, oval; Carniola, Carinthia, Styria, Neuchâtel. *S. decumbens*, Koch; flowers white, in corymbose cymes, stem decumbent, leaves simple, serrate towards the tip; Carniola, Lombardy, rare.

Tribe RUBEÆ.—Calyx persistent, inferior; stamens very numerous; carpels numerous, each ripening into a small drupe (drupel).

3. RUBUS, L.

Flowers in terminal or axillary corymbose panicles, rarely solitary, white or pink; stipules persistent, adnate to the leaf-stalk; creeping or climbing shrubs, mostly prickly.

R. saxatilis, L., is the only alpine species; flowers small, white, few, fruit red, of only 2–3 drupels, leaves of 3 leaflets, stem erect, 6–18 in.; mountain woods, common. *R. Idæus*, L., Raspberry, is also common in woods; and *R. cæsius*, Dewberry, occurs frequently in hedges. Of *R. fruticosus*, L., Blackberry, Bramble, up-wards of sixty sub-species are recorded by Gremli as growing in Switzerland alone; the minute characters which distinguish them would exceed the limits of this work.

Tribe POTENTILLEÆ.—Calyx inferior, persistent, in two whorls; carpels numerous, each ripening into an achene. Genera 4–9.

4. DRYAS, L.

Leaves simple, with adnate stipules; flowers large, solitary; calyx-lobes 8–9; petals 8–9; seed-vessel termi-nated by a feathery style.

D. octopetala, L. (Pl. 31); flowers 1–1½ in. diameter, white, calyx covered with black glandular hairs, leaves oblong-ovate, crenate, covered with a white wool on the under side, stem prostrate, fruit very feathery. This alpine plant, very striking both in flower and in fruit, is frequent in high pastures in the Alps, Jura, and Pyrenees.

5. GEUM, L.

Radical leaves pinnate, upper nearly simple, all with adnate stipules; flowers yellow; calyx 5-lobed, with 5 outer sepals; petals 5; carpels numerous, each ending, in fruit, in a persistent stiff style, often hooked at the tip.

G. urbanum, L., Wood-Avens, with small yellow flowers, is very common in woods and hedges. *G. rivale*, L., with larger flowers, and calyx large, red-brown, is frequent in wet places. The following are alpine :—*G. pyrenaicum*, L.; petals considerably longer than calyx, carpels small, terminal lobe of leaf orbicular-reniform; Pyrenees. *G. montanum*, L. (*Sicversia montana*, Spr.); stem without runners, flower solitary, leaf - segments unequally crenate, terminal segment very large; high pastures; Alps, Jura, Pyrenees. *G. reptans*, L. (*Sieversia reptans*, Spr.); flower large, solitary, stem with long runners, leaf-segments sharply dentate, terminal segment 3–5-cleft; high; Alps, Dauphiny, Pyrenees. *G. intermedium*, Ehrb., and *inclinatum*, Schleich., are probably hybrids.

6. COMARUM, L.

Resembling *Geum*, but fruit not awned ; flowers purplish-brown. Not alpine.

C. palustre, L.; petals dark purple, smaller than the purplish sepals ; wet places ; Switzerland, Pyrenees, local.

7. SIBBALDIA, L.

Flowers in terminal cymes, small, yellow; sepals, outer sepals, and petals 5–7; stamens 4–10; leaves of 3 leaflets.

S. procumbens, L.; stem procumbent, leaflets wedge-shaped, 3 - toothed at the tip; high pastures; Alps, Vosges, Pyrenees.

8. FRAGARIA, L.

Leaves usually of 3 leaflets, with adnate stipules; flowers usually white; calyx-lobes 5, with 5 outer sepals; carpels very numerous, ultimately separated on the surface of the very fleshy receptacle.

F. vesca, L., Wild Strawberry; calyx spreading horizontally or reflexed in fruit; common. *F. elatior*, Ehrh., Hautboy Strawberry; flowers larger, calyx spreading horizontally in fruit. *F. collina*, Ehrh., Alpine Strawberry; calyx erect in fruit, flowers often greenish; hills and woods.

9. POTENTILLA, L.

Leaves compound, with adnate stipules; flowers usually white or yellow, solitary or in corymbose cymes; calyx-lobes, outer sepals, and petals usually 5; carpels numerous, on a dry receptacle.

A. Shrubby; flowers yellow :—*P. fruticosa*, L.; flowers large, in few-flowered cymes, leaves pinnate, very silky; Pyrenees, local.

B. Herbaceous (as the remaining sections); flowers yellow, sepals and petals usually 4 each :—*P. Tormentilla*, Scop. (*Tormentilla erecta*, L.); flowers small, stem prostrate; very common. *P. procumbens*, Sibth.; larger, leaflets 3 or 5; Pyrenees.

C. Flowers white, sepals and petals 5 each :—*P. Fragariastrum*, Ehrh., Barren Strawberry; flowers very small, leaves of 3 leaflets, very silky beneath; hedge-banks, very common. *P. micrantha*, Ram., resembling the last, but flowers sometimes pinkish, stem-leaves simple, no stolons; hedge-banks. *P. splendens*, Ram. ; flowers large, solitary, on long stalks, leaves covered beneath with silky hairs; Pyrenees. The following are alpine or sub-alpine :—*P. caulescens*, L. ; flowers large, stem many-flowered, leaflets 5–7, nearly sessile, silky, ciliate, filaments villous ; rocky ; Alps, Jura, Dauphiny, Pyrenees. *P. petiolulata*, Gaud. ; resembling the last, but less silky, leaflets stalked ; rare ; Salève. *P. Clusiana*, Jacq. ; stem usually 3-flowered, leaves woolly beneath, filaments glabrous ; rare ; Tirol, Carinthia, Styria. *P. rupestris*, L. ; radical leaves pinnate, upper of 3 leaflets, flowers numerous, in corymbose cymes, stem 12–18 in. ; Switzerland, Jura, Pyrenees. *P. grammopetala*, Mor. ; leaves all of 3 leaflets, petals narrow, plant viscous-villous ; very rare ; Grisons. *P. alba*, L. ; radical leaves pinnate, stem-leaves few, of 3 leaflets or simple, flowers large, in 1–5-flowered cymes ; Switzerland (rare), Jura, Dauphiny, Pyrenees. *P. nivalis*, Lap. ; flowers numerous, large, dirty white, all the leaves pinnate or digitate, plant very hairy ; high ; Pyrenees, Dauphiny, *P. alchemilloides*, Lap. ; resembling the last, but still more silky, stipules acuminate, amplexicaul ; Pyrenees.

D. Flowers pink :—*P. nitida*, L. (Pl. 32) ; plant cæspitose, stem usually 1-flowered, few-leaved, radical leaves of 3 leaflets, covered with silky hairs ; high ; Jura, Carniola, Carinthia, Tirol, Dauphiny, Pyrenees.

E. Flowers yellow ; sepals and petals 5 each ; leaves all of 3 leaflets ; alpine plants :—*P. nivea*, L. ; leaves tomentose, snow-white beneath from a coat of silky hairs ; very high ; Southern Switzerland, Tirol, Dauphiny, rare. *P. grandiflora*, L. ; flowers large, stem erect, 4–8 in., leaflets

silky beneath; Switzerland, Styria, Tirol, Dauphiny, Pyrenees, rare. *P. minima*, Hall.; flowers very small, stem 1-2 in., 1-2-flowered, leaves bright green, glabrous above; high; Alps, frequent. *P. frigida*, Vill.; flowers small, stem 1-2 in. high, 1-2-flowered, leaves dull green, villous on both sides with viscid hairs; high; Alps, frequent. *P. norvegica*, L.; stem 8-10 in., flowers large, in crowded terminal cymes, hirsute; Salzburg.

F. Flowers yellow; sepals and petals 5 each; leaves all digitate or pinnate:—*P. anserina*, L.; Silver Weed, with interruptedly pinnate very silky leaves, is a very common wayside weed. *P. cinerea*, Chaix., with very silky digitate leaves; *P. opaca*, L., with more deeply-toothed leaflets, and plant tinted red; *P. verna*, L. (including *Gaudini*, Grml., *aurulenta*, Grml., and *prostrata*, Grml.), leaflets obovate or cuneiform, toothed only in their upper part; grassy places, common; are lowland plants. The following are alpine or sub-alpine:—*P. multifida*, L.; stem 6-12 in., much branched, segments of leaflets linear, leaves tomentose beneath; Zermatt, Dauphiny, rare. *P. aurea*, L.; flowers orange-yellow, stem single, 3-12 in., 3-5-flowered, leaflets with a fringe of silky hairs; alpine pastures. *P. alpina*, Willk.; resembling the last, but stems cæspitose, 1-flowered; alpine pastures. *P. alpestris*, Hall (*villosa*, Crntz., *salisburgensis*, Hænk., *sabauda*, DC.); resembling *P. verna*, but flowers larger, of a brighter yellow, leaflets more deeply incised, and with auriculate stipules; alpine pastures, common. *P. heptaphylla*, Mill.; flowers large, stem erect, many-flowered, leaflets 7-9, dentate nearly throughout; pastures, local. *P. pyrenaica*, Ram.; resembling *P. alpestris*, but a larger stouter plant, with more leafy stem, and flowers on shorter

stalks; Pyrenees. *P. intermedia*, L.; flowers on long
stalks, upper leaves opposite; Pyrenees, Dauphiny. *P.
delphinensis*, G. and G.; stem suffruticose, stipules lanceolate-acuminate; Dauphiny.

G. Flowers yellow, sepals and petals 5 each, leaves
digitate, with usually 5 leaflets; not alpine. *P. reptans*,
L. (including *ascendens*, Grml.), is a very common road-
side plant.

H. Flowers yellow, sepals and petals 5 each, floral
stem terminal; not alpine. *P. argentea*, L.; leaflets
usually 5, incised, very white beneath; walls and dry
places in the lowlands. *P. collina*, Koch (*Güntheri*,
Pohl); resembling the last, but leaflets grey-tomentose
beneath; rare. *P. cinerea*, auct. (*incana*, Fl. Wett.);
similar, but leaves grey-tomentose on both sides; rare.
P. canescens, Bess. (*inclinata*, Vill.); stem erect, leaflets
oblong-lanceolate; occasional. *P. recta*, L.; stem erect,
very leafy, 12–18 in., flowers sulphur-coloured; dry
banks, local. *P. supina*, L.; petals shorter than the
calyx, stem decumbent, flowers small, pale yellow; Bâle.

Tribe POTERIEÆ.—Petals 4, 5, or 0; carpels 1–3,
developing into achenes enclosed within the calyx-tube.
Genera 10–13.

10. AGRIMONIA, L.

Calyx 5-lobed; petals 5; flowers yellow, in terminal
spike-like racemes; stamens 12–20; carpels 2; leaves
pinnate. Not alpine.

A. Eupatoria, L., Agrimony; calyx-tube obconical;
very common in hedge-banks and thickets. *A. odorata*,

Mill.; a scented plant, with larger flowers and campanulate calyx-tube; rare; Bâle, Tirol, Carniola.

11. AREMONIA, Neck.

Calyx 5-toothed; petals 5; stamens 5–10; teeth of
calyx ultimately forming a spiny ring.

A. agrimonioides, DC.; flowers yellow, stem 3–5 in.,
stem-leaves of 3 leaflets, leaflets roundish-ovate; Tirol,
Styria, Carniola.

12. ALCHEMILLA, L.

Calyx 4–5-lobed, with 4–5 outer sepals; petals 0;
flowers small, green; stamens few; carpels 1–5. Prostrate herbs.

A. vulgaris, L., Lady's Mantle, with large, nearly
orbicular or reniform dentate soft leaves; very common
in damp meadows; and *A. arvensis*, L., with fan-shaped
leaves and very minute flowers; in cultivated land. The
remainder are sub-alpine :—

A. Radical leaves roundish-reniform, 7–9-lobed :—*A.*
montana, Willd.; radical leaves pubescent, divided to
about ⅓, serrate, all the leaves silky; alpine pastures;
Switzerland, frequent. *A. pubescens*, Bieb.; resembling
the last, but the lobes almost truncate, serrate only near
the tip; local. *A. fissa*, Schum. (*pyrenaica*, Duf.); radical
leaves nearly glabrous, more deeply divided, lobes obovate, stem glabrous; damp alpine rocks, local; Jura,
Pyrenees, Carpathians.

B. Radical leaves divided nearly to the base into 5–7
segments :—*A. subsericea*, Reut.; leaves densely hairy

beneath, greyish-white. *A. alpina*, L. (including *Hoppeana*, Rchb.); leaf-segments narrow, leaves shining with silky hairs; a pretty and very common alpine plant. *A. pentaphyllea*, L. (including *cuneata*, Gaud.); leaves only slightly hairy, the three middle segments deeply incised; very high; Switzerland, Tirol, Pyrenees.

13. POTERIUM, L.

Calyx 4-lobed, coloured; petals 0; stamens numerous; carpels 1–3. Not alpine.

P. Sanguisorba, L. (*dictyocarpum*, Spach), Smaller Burnet; *P. muricatum*, Spach; and *P. officinale*, Hook. (*Sanguisorba officinalis*, L.), Greater Burnet; distinguished from one another by very minute characters; are abundant in damp lowland meadows.

Tribe ROSEÆ.—Calyx 5-lobed; petals 5; stamens numerous; carpels numerous, developing into achenes enclosed within the fleshy calyx-tube.

14. ROSA, L.

Flowers large, handsome; stem often prickly; leaves pinnate, with stipules adnate to the leaf-stalk; calyx-tube ultimately coriaceous or very fleshy, brightly coloured, and enclosing the achenes.

But few species of Rose are alpine. The most readily distinguished of these are:—*R. cinnamomea*, L., in which the branches are reddish-brown, and the stipules of the barren branches have erect edges, almost conniving in a tube, the prickles in pairs beneath the stipules; wet places; Engadine, Valais, Winterthur, &c. *R. alpina*, L.

(Pl. 33), with numerous sub-species; a small erect shrub, corolla bright red, the flowering branches usually without spines, leaflets 7–11, with scattered glands, fruit red, usually ovoid-oblong; alpine, frequent. *R. rubella*, Sm.; resembling the last, but with pale pink or white petals, fruit red; Jura. *R. spinosissima*, L., Scotch Rose; a small erect shrub, with globular nearly black fruit, the stem densely covered with small thorns; Jura. *R. pomifera*, Herrm.; sepals and flower-stalks very glandular, sepals pinnatifid, fruit large, usually globular; Alps, Jura, Styria, rare. *R. mollis*, Sm.; resembling the last, but the glands weaker; sub-alpine woods. *R. ferruginea*, Vill.; sepals usually entire, petals bright red, fruit small, globular, stem almost without prickles; Alps, Jura. *R. montana*, Chaix; leaflets small, nearly round, obtuse, calyx-tube and flower-stalk covered with glandular thorns; mountain woods, rare.

The number of Swiss species of *Rosa* given by Gremli is about fifty, many of them very difficult to distinguish, and possibly hybrids. They may be arranged under the following groups (after Hooker):—*R. spinosissima*, L.; small erect shrubs, with crowded prickles and small nearly globular fruit. *R. villosa*, L.; with long prickly branches, leaflets very hairy beneath, densely glandular sepals, and prickly globose or turbinate fruit. *R. rubiginosa*, L., Sweet Briar; flower-stalks covered with a glandular pubescence, leaflets densely glandular; fragrant. *R. canina*, L., Dog-Rose; with long prickly branches, sepals glabrous, pinnate, reflexed, leaflets nearly glabrous, fruit not prickly. *R. arvensis*, Huds.; flowers few, pale, sepals deciduous, styles united into an exserted column.

Tribe POMEÆ.—Calyx 5-toothed; petals 5; stamens numerous; carpels 2–5, each with two ovules; fruit a pome (pseudocarp) or drupe. Trees or shrubs, usually with simple leaves and deciduous stipules. Genera 15–20.

15. PYRUS, L.

Leaves simple; fruit (pseudocarp) a pome. Not alpine. *P. communis*, L., Wild Pear; and *P. Malus*, L., Crab-Apple; woods and hedges.

16. SORBUS, L.

Flowers in compound corymbose cymes; fruit a small 2–5-celled pome; leaves simple or compound.

S. Aucuparia, L., Rowan, Mountain Ash; leaves pinnate, fruit small, globular, scarlet; very common. *S. domestica*, L.; leaves pinnate, fruit yellowish-red, pear-shaped; woods, not common. *S. Aria*, Crntz.; flowers white, leaves simple or lobed, very white underneath; rocky woods. *S. torminalis*, L., Service-Tree; flowers white, leaves 6–10-lobed, serrate, not white beneath when full-grown, fruit pear-shaped, greenish-brown; woods. *S. scandica*, Fr.; flowers white, leaves inciso-pinnatifid, slightly tomentose beneath; Jura, Pyrenees. *S. chamæmespilus*, Crntz.; flowers pink, in dense corymbs, leaves simple, elliptical, serrate, green beneath; rocky places; Alps, Jura, Vosges, Pyrenees. *S. Hostii*, Jacq.; resembling the last, but leaves tomentose beneath; Alps, Jura, rare.

17. MESPILUS, L.

Flowers large, solitary; fruit with a bony endocarp. *M. germanica*, L., Medlar; thickets, rare.

XXXIV.—SAXIFRAGA OPPOSITIFOLIA.

18. CRATÆGUS, L.

Flowers small; leaves simple, lobed or pinnatifid; fruit a 1–5-celled drupe; branches ending in a spine.

C. oxyacantha, L., Hawthorn, Whitethorn; carpels 2–3; common. *C. monogyna*, Jacq.; carpel 1; not so common.

19. COTONEASTER, L.

Flowers small; leaves coriaceous, simple; fruit a small pome with 2–5 1-seeded stones.

C. vulgaris, Lindl.; flowers 1–5, calyx nearly glabrous; rocky mountain slopes, not uncommon. *C. tomentosa*, Lindl.; flowers 3–8, calyx downy; rocky places, local.

20. ARONIA, Pers.

Flowers in erect racemes, white; fruit a 5-celled 3–5-seeded pome.

A. rotundifolia, Pers. (*Amelanchier vulgaris*, Mœnch.), Snowy Medlar; leaves ovate, obtuse, downy beneath, ultimately glabrous, fruit blue-black; rocky mountain slopes, local; Alps, Jura, Pyrenees.

Order XXX.—SAXIFRAGACEÆ.

Flowers regular; calyx 5-lobed; petals 5; stamens usually 10; ovary usually half-inferior and 2-celled; fruit a 2–4-celled capsule. A large order, almost entirely confined to the Arctic and Northern Temperate Zones.

1. SAXIFRAGA, L.

Flowers in cymes, rarely solitary; calyx-tube free or partially adnate to the ovary; ovary 2-lobed, 2-celled; styles 2; capsule 2-valved, 2-beaked.

Only a very few species of Saxifrage grow in the lowlands, viz.:—*S. tridactylites*, L.; a small annual plant 2–4 in. high, with small white flowers in many-flowered cymes, and cuneate 3–5-fid leaves; walls and dry places. *S. granulata*, L.; stem 6–18 in., with bulbs in the axils of the lower leaves, flowers ½ in. diam., white, drooping, leaves palmately lobed; meadows. *S. bulbifera*, L.; all the stem-leaves bulbiferous; Western Switzerland. All the remaining species are alpine.

A. Flowers pink or violet :—*S. oppositifolia*, L. (Pl. 34); flowers sessile, solitary, ½ in., violet, stem prostrate, leaves small, ciliate, very crowded, opposite, with a pore at the tip; one of the most beautiful alpine plants; wet rocks, not uncommon; Alps, Jura, Pyrenees. *S. biflora*, All.; flowers 2 or more, petals lanceolate, leaves opposite, not so crowded, ciliate; very high; Alps, Dauphiny. *S. retusa*, Gou.; flowers small, stalked, light purple, calyx-teeth not ciliate, stem-leaves opposite, with 3–5 pits at the tip; very rare; Monte Rosa, Simplon, Styria, Pyrenees. *S. Rudolphiana*, Hornsch.; flowers solitary, terminal, pink, calyx-teeth and leaves glandular-ciliate, plant cæspitose; high; Switzerland, Carniola, local. *S. macropetala*, Kern. (*Kochii*, Horn.); flowers large, 2 or more, lilac, petals lanceolate; Switzerland, Tirol, Carniola, local. *S. media*, Gou.; flowers 3–7 in a simple cyme, petals pink, shorter than the very dark purple calyx, leaves oblong-spathulate, glaucous, with mem-

XXXV.—SAXIFRAGA MUTATA.

XXXVI.—SAXIFRAGA AMBIGUA.

branous margin and a row of calcareous pits; Pyrenees. *S. atropurpurea*, Sternb.; stem cæspitose, many-flowered, leaves of barren shoots on long stalks, 3–5-cleft, petals obovate-elliptical, dark purple; very high, local.

B. Flowers yellow:—*S. mutata*, L. (Pl. 35); radical leaves in a dense rosette, fleshy, ligulate, with a marginal row of calcareous pits, stem 2–4 in., branched, petals linear, orange-yellow; Southern Switzerland, Jura, Pyrenees. *S. aretioides*, Lap.; petals oblong, denticulate, leaves linear, with a few pits; Pyrenees. *S. ambigua*, DC. (Pl. 36); Pyrenees; a very remarkable plant, is possibly a hybrid between *S. media* and *aretioides*. *S. Hirculus*, L.; petals golden yellow, dotted with red and tubercled at the base, sepals free, reflexed, ciliate, leaves linear, not pitted; Jura, rare. *S. muscoides*, Wulf.; stem densely cæspitose, flowers citron or pink, leaves linear, narrowed into the leaf-stalk; high; Northern Switzerland, Jura, Pyrenees. *S. moschata*, Wulf.; resembling the last, but plant covered with viscid hairs. *S. Allionii*, Gaud.; similar, but petals greenish-yellow or white, obovate; rare. *S. crocea*, Gaud.; similar, but petals saffron-yellow. *S. Facchinii*, Koch; stem 1–3-flowered, petals as broad as and longer than the calyx-teeth; rare. *S. sedoides*, L.; plant loosely cæspitose, fertile stem leafless, petals lanceolate, acute, shorter and narrower than the calyx-teeth; Alps, Pyrenees. *S. Hohenwartii*, Sternb.; fertile stem leafy, petals as broad as calyx-teeth, lemon-yellow with dark purple apex; Tirol, Carinthia, Styria, Carniola. *S. stenopetala*, Gaud.; stem 1-flowered, petals lemon-yellow, linear, lower leaves pinnatifid; Eastern Switzerland. *S. aizoides*, L. (Pl. 37); calyx half-inferior, leaves linear, ciliate; moist; very common. *S. arachnoides*, Sternb.; stem prostrate, plant covered with long white arachnoid hairs, leaves all radical; Tirol, rare.

C. Flowers yellowish-white or greenish-white :—*S. hieraciifolia*, W.K.; flowers in a dense spike, light green with purple edge, bracts leaf-like, leaves all radical, ovate-lanceolate, entire; Styria, rare. *S. bryoides*, L.; stem 1-2 in. high, 1-2-flowered, leaves linear-lanceolate, with foliaceous buds in their axils; high; Alps. *S. aspera*, L.; a larger plant, with more numerous flowers, and the foliaceous buds larger; high; Alps, Pyrenees. *S. Seguieri*, Spreng.; cæspitose, leaves spathulate, stem usually 1-flowered, downy, petals oblong-linear, as long as calyx-teeth; glaciers, rare; Switzerland, Tirol. *S. pygmæa*, Haw.; densely cæspitose, leaves fleshy, entire, those of the barren shoots linear, petals oval; high; local. *S. Aizoon*, L. (Pl. 38); stem branched, 4-6 in., glandular-hairy, petals green or white, radical leaves in rosettes, thick, glabrous, ligulate, bordered with a row of pits; rocky places; Alps, Jura, Vosges.

D. Flowers white, with red or yellow spots, or otherwise variegated :—*S. umbrosa*, L., London Pride; stem 6-12 in., branched, leafless, radical leaves nearly orbicular, narrowed into the leaf-stalk, crenate, flowers small, petals with red and yellow spots; Pyrenees. *S. hirsuta*, L. (including *Geum*, L.); more hairy, leaves more sharply toothed; Pyrenees. *S. rotundifolia*, L.; radical leaves cordate-reniform, stem-leaves not cordate, shortly stalked, stem 8-12 in., much branched, petals narrow, spotted; woods, common. *S. cuneifolia*, L. (including *subintegra*, Ser.); leaves glabrous, cuneate, crowded, thick, flowering stem branched, leafless, petals usually with one yellow

XXXVII.—SAXIFRAGA AIZOIDES.

XXXVIII.—SAXIFRAGA AIZOON.

spot; high, moist; Switzerland, Jura, Pyrenees, rare.
S. Cotyledon, L.; stem 1–2 ft., much branched, leafy, radical and stem leaves oblong - obovate, glabrous, cartila-
ginous, with very thick margin, serrate, and furnished with
a row of calcareous pits, petals wedge-shaped, red at the
base or spotted; high, moist; Switzerland, Styria, Carinthia, Pyrenees, rare. *S. stellaris*, L.; stem 1–2 in.,
leafless, flowers small, radical leaves obovate-cuneate,
3-lobed at the apex, petals unequal; very common. *S.
glabrata*, Sternb.; similar, but very glabrous, leaves
smaller; high, local. *S. Hostii*, Tausch. (including
rhætica, Kern.); stem-leaves with a marginal row of
calcareous pits, leaves of rosettes ligulate with cartila-
ginous margin, crenate; Tirol, Carinthia, Carniola, local.
S. purpurata, Gaud.; similar, but stem and branches purple,
petals with red veins; Switzerland, rare (Splügen,
Devil's Bridge). *S. maculata*, Rchb.; densely cæspitose,
leaves cuneate, petals white with long purple spots;
Switzerland, rare (St. Gothard, Rhone glacier).

E. Flowers usually pure white; leaves with calcareous
pits; stem-leaves alternate :—*S. crustata*, Vost.; rosette-
leaves furrowed, entire, linear, covered with a strong
incrustation; Tirol, Carniola, Carpathians. *S. altissima*,
Kern.; stem 3–6 in., glandular-hairy, radical leaves ligulate, serrate, flowers in a corymbose panicle, sometimes
spotted, flower-stalk glandular; Styria. *S. diapensoides*,
Bell.; stem densely viscid-glandular, 2–3-flowered, petals
obovate-cuneate, milk-white, leaves linear-oblong; Switzerland, Tirol, Dauphiny, local. *S. cæsia*, L.; rosette-
leaves densely crowded, recurved from the base, petals
obovate, obtuse; high, frequent. *S. valdensis*, DC.; a
smaller plant, with smaller flowers, stem glandular;

Savoy, Mont Cenis, rare. *S. squarrosa*, L.; stem 1–3
in., few-flowered, glandular-hairy, rosette-leaves erect,
blunt, flowers sometimes yellowish; Tirol, Carniola,
frequent. *S. Burseriana*, L.; very cæspitose, stem
usually 1-flowered, glandular-hairy, flowers large, white,
sometimes with red veins, petals crenulate, calyx half-
superior, reddish, leaves subulate, ending in a sharp
spine; dry places; Alps. *S. Vandellii*, Sternb.; very
cæspitose, stem 2–9-flowered, densely glandular-hairy,
flowers large, milk-white, petals obovate-cuneate, much
longer than the calyx, leaves linear-oblong, mucronate; Switzerland, Tirol, rare. *S. tombeanensis*, Boiss.;
densely cæspitose, stem 1–3-flowered, glandular-hairy,
flowers very large, petals obovate-cuneate, 3–4 times as
long as calyx-teeth, leaves ovate-lanceolate, ending in a
soft spine; Tirol, rare (Monte Baldo). *S. longifolia*,
Lap.; a very remarkable and distinct plant, with the
flowers in a very large pyramidal panicle, 6–18 in. high,
radical leaves very numerous, linear, coriaceous; Pyrenees. *S. lingulata*, Bell.; resembling the last, but a more
glabrous plant, with narrow petals and fewer flowers;
Pyrenees. *S. Lantoscuna*, Boiss.; also an allied species,
but with shorter and broader leaves and fewer flowers;
Maritime Alps.

F. Flowers usually white; leaves without calcareous
pits :—*S. controversa*, Sternb. (*petræa*, Gaud.; *Bellardii*,
All.); a small annual plant resembling *tridactylites*, but
more robust, and with somewhat larger flowers; Alps,
Dauphiny, Pyrenees. *S. cernua*, L.; flowers usually
solitary, drooping, stem leafy, leaves all palmately 5–7-
lobed, the upper ones bearing bulbs in their axils. *S.
geranioides*, L.; flowers large, pure white, numerous,

tubular, calyx half-superior, radical leaves cordate, nearly orbicular in outline, with lanceolate lobes; Eastern Pyrenees, rare. *S. petræa*, L.; stem decumbent, leafy, hairy, flowers large, leaves palmately 3 - cleft, lobes inciso-dentate, petals obovate, twice as long as calyx; Southern Tirol, Carpathians, rare. *S. pedemontana*, All.; flowers large, 2–12, radical leaves pinnatifid, petals narrowed to a distinct claw, calyx-teeth linear; very rare; Monte Rosa. *S. pentadactylis*, Lap.; flowers 5–9, leaves flabelliform-pinnatifid; Eastern Pyrenees. *S. nervosa*, Lap.; flowers 4–12, plant covered with glutinous hairs, and with a balsamic odour; Pyrenees. *S. ascendens*, L.; stem 1–2 ft., only slightly branched, very leafy, leaves 5–7-lobed, fleshy, covered with viscid hairs; Pyrenees. *S. ajugæfolia*, L.; flowers on long axillary 1–3-flowered flower-stalks, leaves 2–5-lobed, nearly glabrous; Pyrenees. *S. pubescens*, Pourr.; flowers 1–5, milk-white, stem about 1 ft., viscid, radical leaves with linear segments, stem-leaves linear; Pyrenees. *S. grœnlandica*, L.; flowers 5–9, petals broad, stem 4–6 in., glutinous-pubescent; resembling *muscoides*, but with larger flowers; Pyrenees. *S. exarata*, Vill.; flowers 3–9, leaves cuneate, narrowed at the base, 3-cleft with projecting lobes, petals twice as long as calyx-teeth, sometimes pale yellow or spotted; Alps, Tirol, Carniola, Carinthia, Dauphiny. *S. intricata*, Lap.; resembling the last, but with a slenderer stem and smaller milk-white flowers; Pyrenees. *S. androsacea*, L.; cæspitose, pubescent, petals obovate, emarginate, twice as long and broad as the calyx-teeth, stem 1–2 in., 1–3-flowered, radical and stem-leaves lanceolate, the former often 3-toothed at the apex; high, moist, frequent. *S. tenella*, Wulf.; stem 2–4 in., branched, calyx-teeth cuspidate, leaves linear-subulate, cuspidate, stem-leaves with axillary buds; Styria; rare. *S. plani-folia*, Lap.; densely cæspitose, flowers 1–5, petals obovate, emarginate, twice as long as calyx-teeth, leaves linear-oblong, quite entire; glaciers; Alps, Pyrenees, rare. *S. sponhemica*, Gmel.; flowers 3–9, calyx-teeth acuminate, leaves linear, acuminate, aristate, plant slightly hairy; Jura. *S. hypnoides*, L.; flowers 5–9, calyx-teeth mucronate, leaves linear, acuminate, mucronate, plant cæspitose, hairy; frequent.

Several other species are described in Floras of Switzerland and Pyrenees; but they are either slight varieties, or sometimes probably hybrids.

2. PARNASSIA, L.

Leaves entire, mostly radical; flowers large, solitary; stamens alternating with 5 large fan-shaped glandular scales; styles 3–4; capsule 3–4-valved.

P. palustris, L., Grass of Parnassus (Pl. 39); flower white, stem with one ovate-cordate leaf; damp places, common. *P. alpina*, Brügg., is only a small mountain variety.

3. ZAHLBRUCKNERA, Rchb.

Flowers small, solitary, green; ovary 2-lobed; capsule 2-celled.

Z. paradoxa, Rchb.; stem decumbent, lower leaves cordate-reniform, 5–7-lobed, on long stalks, flowers on long stalks, petals small, green, shorter than the calyx-teeth; moist rocks; Styria, Carinthia, Southern Tirol, Lombardy, very rare.

XXXIX.—PARNASSIA PALUSTRIS.

.

4. CHRYSOSPLENIUM, L.

Flowers very small, green or yellow, in axillary or terminal cymes; petals 0; stamens 8–10; ovary inferior, 1-celled; capsule 2-lobed. Not alpine.

The two species of Golden Saxifrage, *C. alternifolium*, L., with alternate, and *oppositifolium*, L., with opposite leaves, are found in wet places, by springs, &c., blossoming very early in the spring; the former much the most common.

Order XXXI.—GROSSULARIEÆ.

Flowers regular; calyx usually 5-lobed, persistent; petals 5; ovary 1-celled, entirely inferior, with parietal placentation; fruit an indehiscent berry. A small order, chiefly European.

1. RIBES, L.

Flowers small, solitary or in racemes; styles 2. Shrubs.

R. Grossularia, L., Gooseberry (usually the form *R. uva-crispa*, L., with small smooth berries); *R. nigrum*, L., Black Currant; *R. rubrum*, L., Red and White Currant; are widely distributed, though not very common. *R. alpinum*, L.; flowers unisexual, in erect racemes, yellowish-green; bushy places; Alps, Pyrenees, local. *R. petræum*, Wulf.; flowers bisexual, reddish, in racemes which are ultimately pendant; bushy places; Switzerland, Jura, Styria, Vosges, local.

Order XXXII.—CRASSULACEÆ.

Flowers regular, in terminal or axillary cymes; sepals and petals usually 5; stamens usually 10; carpels usually 5, distinct, maturing into many-seeded follicles. Herbs or small shrubs with succulent leaves. A rather large order, distributed chiefly through the colder and temperate regions.

1. RHODIOLA, L.

Flowers unisexual; sepals 4; petals 4 or 0; stamens 8.
R. rosea, L. (*Sedum Rhodiola*, DC.), Rose-root; stem erect, leafy, flowers small, reddish-yellow, leaves fleshy, ovate-lanceolate, serrate towards the tip; moist rocks; Switzerland, Jura, Vosges, Pyrenees.

2. SEDUM, L.

Flowers bisexual, in unilateral cymes, star-like; sepals and petals usually 5; stamens usually 10; carpels usually 5, distinct or slightly connate at the base. Very succulent, and (except Sections A., B.) all wall or rock plants.

A. Leaves flat, succulent; stem erect or decumbent; flowers very numerous; stamens 10:—*S. Anacampseros*, L.; flowers light purple or pink, stems numerous, decumbent, leaves obovate, entire; Western and Southern Switzerland, Tirol, Dauphiny, Pyrenees. *S. maximum*, Sut.; flowers yellowish, petals somewhat hooded, stems numerous, erect, leaves cordate-oblong, dentate, the upper ones often opposite or in whorls of three; stony thickets; Switzerland, Jura, Dauphiny. *S. Telephium*, L. (*purpurascens*, Koch), Orpine; flowers pink or purple, stem

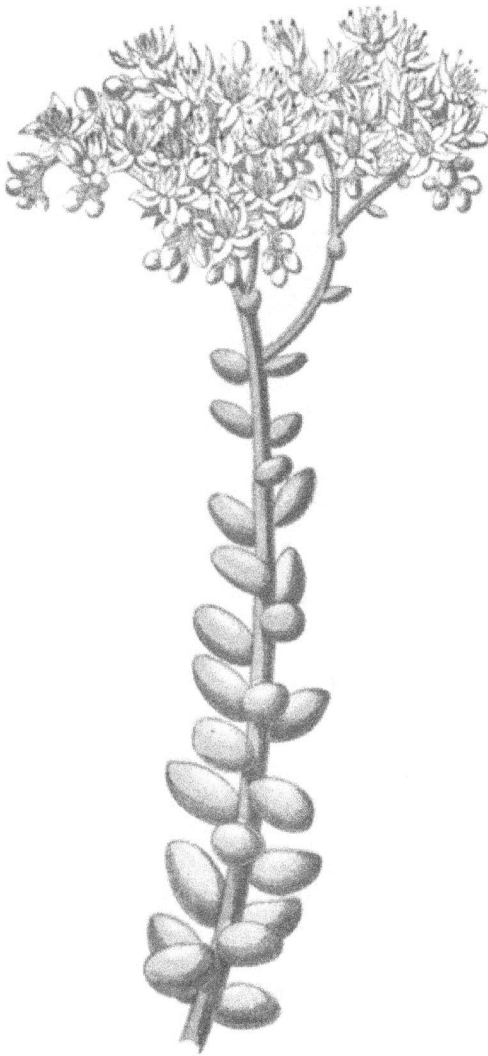

XI.—SEDUM TURGIDUM.

ascending, leaves alternate, obscurely toothed; banks;
Switzerland, Jura, Pyrenees. *S. Fabaria*, Koch; very
similar to the last, more slender, the leaves more wedge-
shaped; Jura, Vosges, Dauphiny, Pyrenees, local. *S.
Cepæa*, L.; flowers pink, stalked, forming a branched
panicle, petals acuminate, stem simple, slender, decumbent then ascending, leaves spathulate, entire, the lower
ones often in whorls; stony places; Southern Switzerland, Jura, Tirol, Pyrenees. *S. stellatum*, L.; flowers
pink, sessile, in compound scorpioid spikes, stem simple
or branched, leaves dentate, sometimes opposite or in
whorls; Pyrenees.

  B. Leaves cylindrical; stamens 5:—*S. rubens*, L.;
flowers light pink, sessile, in glandular-pubescent unilateral spikes, stem tinged with red; vineyards; Switzerland (rare), Jura, Tirol, Vosges, Pyrenees.

  C. Leaves cylindrical; petals 6; stamens 12:—*S. his-
panicum*, L.; flowers nearly white, leaves bluish-green,
capsule pubescent; rocks; Central and Southern Switzerland, rare.

  D. Leaves cylindrical; petals 5; stamens 10; plant
without barren shoots:—*S. villosum*, L.; petals pink
with a dark central line, plant covered with a glandular
pubescence; moist; Switzerland, Jura, Vosges, Dauphiny,
Pyrenees. *S. atratum*, L.; petals white or greenish-
yellow, calyx reddish-brown, flowers in compact cymes,
leaves thickened upwards, plant glabrous, tinged with
reddish-brown; high; Switzerland, Jura, Pyrenees. *S.
annuum*, L.; flowers yellow, in elongated spike-like
cymes, nearly sessile, leaves not thickened upwards, plant
glabrous; Switzerland, Vosges, Pyrenees.

  E. Leaves cylindrical; petals 5; stamens 10; plant

with cæspitose barren shoots; flowers white or pink:—
S. album, L., White Stonecrop; flowers white, inflorescence much branched, leaves alternate, linear; rocks and walls, common. *S. dasyphyllum*, L.; flowers white, leaves short and thick, often opposite and spotted with red; rocks and walls, frequent. *S. hirsutum*, All.; petals pink with a darker central line, or white, acuminate, leaves distant, hairy, plant glandular-pubescent; Pyrenees. *S. turgidum*, Ram. (*micrantum*, G. and G.), (Pl. 40); flowers light pink or white, leaves numerous on the fertile branches, oval-cylindrical; Dauphiny, Pyrenees. *S. anglicum*, Huds.; flowers white or light pink, petals lanceolate, acuminate, keeled, leaves gibbous at the base; Pyrenees. *S. brevifolium*, DC.; flowers white or light pink, stalked, leaves distant, nearly spherical, plant glabrous; Pyrenees.

F. Leaves cylindrical; petals 5; stamens 10; plant with cæspitose barren shoots; flowers yellow:—*S. alpestre*, Vill.; flowers few, pale yellow, petals erect, obtuse, leaves obtuse, stem slender; alpine rocks, frequent. *S. acre*, L., Yellow Stonecrop, Wall Pepper; flowers bright yellow, calyx-teeth gibbous at the base, petals acuminate, leaves very crowded, ovoid-oblong; walls and rocks, common. *S. sexangulare*, L. (Pl. 41); flowers less numerous and somewhat smaller, sepals not gibbous at the base, petals acute, leaves densely crowded, spreading; walls and rocks; Switzerland, Pyrenees, local. *S. reflexum*, L. (*rupestre*, L.), (Pl. 42); flowers bright yellow, usually in large bifurcate cymes, petals lanceolate, spreading, sepals ovate, leaves subulate, acute; stony places; Switzerland, Pyrenees. *S. anopetalum*, DC. (*ochroleucum*, Chaix); petals erect, pale yellow, lanceolate, acuminate, sepals acuminate, leaves mucronate; Southern Switzer-

XLI.—SEDUM SEXANGULARE.

XLII.—SEDUM REFLEXUM.

land, Jura, Pyrenees. *S. altissimum*, Poirr.; flowers very
pale yellow, stem somewhat shrubby, branched, leaves
oval, concave above; Pyrenees, Dauphiny.

3. SEMPERVIVUM, L.

Flowers large, bisexual, in corymbose or panicled cymes;
calyx-teeth and petals 6 or more; stamens usually twice
as many as petals; carpels as many as petals; leaves
fleshy, often covered with arachnoid hairs. All the
species are alpine, and the genus is especially characteristic of the mountains of Switzerland and Tirol. The
species are often difficult to distinguish, and appear to
hybridise readily.

A. Calyx-teeth, petals, and carpels more than 6;
rosette-leaves ciliate at the margin, otherwise glabrous:
—*S. Wulfeni*, Hoppe; petals yellow or whitish, leaves
glaucous, stem and calyx covered with rough glandular
hairs; rare; Engadine, Tirol, Styria, Carniola. *S. alpinum*, G. and S. (*Boutignianum*, G. and G.); petals
linear-lanceolate, three times as long as calyx-teeth,
ciliate, red with a darker central streak, marginal hairs
of rosette-leaves not glandular; rare; Engadine, Ticino,
Tirol, Pyrenees. *S. tectorum*, L., House-Leek; petals
lanceolate, twice as long as calyx-teeth, pink, rosettes
large, rosette-leaves suddenly contracted into a point,
which is usually red, stem-leaves acuminate; rocks;
Switzerland, Tirol, Pyrenees; also in the lowlands cultivated on roofs, &c. *S. acuminatum*, Schott; similar
to the last, but rosette-leaves gradually acuminate, leaves
bluish-grey; Tirol, rare (Botzen). *S. arvenense*, L. and
L.; flowers pink, half the size of the last, petals 1½

times as long as calyx-teeth, stem-leaves keeled, nearly
triangular, stem glandular-pubescent; Puy-de-Dôme.
S. Mettenianum, S. and L.; flowers small, pink, petals
linear, rosettes small, flat, rosette-leaves gradually acuminate, narrowed at the base; Switzerland, Southern
Tirol, rare. *S. dolomiticum*, Facch.; flowers large, dark
red, petals lanceolate, rosettes small, globular, leaf-tips
and stem often tinged with red; Southern Tirol, rare
(Pusterthal, Fassathal). *S. fimbriatum*, S. and L.; flowers
large, light red, petals linear, twice as long as calyx-teeth,
rosette-leaves broadly lanceolate, acute, all the leaves
ciliate with glandular hairs; Southern Tirol, Carinthia,
rare. *S. angustifolium*, Kern.; resembling the last, but
rosette-leaves narrower, petals broader and longer, pink
with a darker central line, filaments glandular-hairy;
Tirol, rare (Oetzthal).

  B. Calyx-teeth, petals, and carpels more than 6; rosette-
leaves with simple and glandular hairs, covering it like
a spider's web:—*S. arachnoideum*, L. (Pl. 43); rosette-
leaves and stem-leaves completely covered by a weft of
arachnoid hairs, flowers pink, petals elliptic-lanceolate,
three times as long as calyx-teeth, stem 4–6 in. high;
alpine rocks, frequent. *S. heterotrichum*, Schott; rosette-
leaves with arachnoid hairs only at the tip, petals pink
with a darker central line, calyx-teeth with a tuft of hairs
at the tip, stem 3–5 in.; Salzburg. *S. Döllianum*, Lehm.;
rosette-leaves and stem-leaves with arachnoid hairs,
which finally disappear, stem 4–6 in., petals twice as long
as calyx-teeth, broadly lanceolate; Switzerland, Carinthia,
rare. *S. Fauconneti*, Reut.; petals longer and broader
than in *arachnoideum*, rosette-leaves warty-glandular
above, with white flexible hairs at the edge; Jura.

XLIII.—SEMPERVIVUM ARACHNOIDEUM.

C. Calyx-teeth, petals, and carpels more than 6; rosette-leaves with simple and glandular hairs, but not arachnoid: —*S. Funckii*, Br.; petals lanceolate, acuminate, pink with a darker central line, rosettes small, globular, cilia of leaves very long; Eastern Switzerland, Tirol, Carinthia, Carpathians, Salzburg. *S. montanum*, L. (Pl. 44); petals lilac with a violet central line, lanceolate, acuminate, four times as long as calyx-teeth, rosettes open, cilia of leaves short; alpine rocks, frequent. *S. debile*, Schott; resembling the last, but stem prostrate, leaves more crowded, petals brownish-red; Western Tirol. *S. rupicolum*, Kern.; petals pale greenish-yellow streaked with red, flowers in trifurcate cymes, rosette-leaves glandular-hairy on the under side only; Tirol (Oetzthal). *S. Gaudini*, Christ (*globuliferum*, Gaud.); flowers very large, in trifurcate cymes, petals linear, yellowish-white, not streaked, three times as long as calyx-teeth, rosette-leaves glandular-hairy on both sides, stem-leaves glandular-ciliate; Valais, Tirol, rare. *S. Pittonii*, S. and K.; flowers small, petals linear-lanceolate, yellowish-white, not streaked, rosettes small, all the leaves glandular-hairy on both sides, tipped with red; Styria, rare. *S. Braunii*, Funck.; petals linear, yellowish-white, twice as long as calyx-teeth, rosettes small, leaves convex on both sides, obtusely angular; rare; Grisons, Carinthia, Gross-Glockner. *S. Widderi*, S. and L.; petals yellowish-white, broadly lanceolate, rosette-leaves flat on the upper, convex on the under side, acutely angled; Tirol, rare.

D. Calyx-teeth, petals, and carpels usually 6; stamens 12; flowers bell-shaped:—*S. hirtum*, L.; petals yellowish white, linear-lanceolate, twice as long as calyx-teeth, fringed, leaves ciliate, rosette-leaves ovate-acute; Tirol,

Carpathians, Erzgebirge. *S. arenarium*, S. and K.; flowers large, petals nearly white, three times as long as calyx-teeth, fringed, rosette-leaves lanceolate; Tirol, Carinthia, Salzburg. *S. Neilreichi*, S. and K.; petals with narrow winged serrated keel, fringed, leaves narrowly lanceolate; Lower Austria. *S. Hildebrandtii*, Schott; keel of petals torn into shreds, leaves lanceolate; Styria.

4. UMBILICUS, DC.

Calyx 5-cleft; corolla gamopetalous, 5-cleft; stamens 10, attached to the corolla.

U. sedoides, DC.; flowers pink, corolla twice as long as calyx, plant tinged with red; rocks; Pyrenees.

Order XXXIII.—HALORAGEÆ.

Flowers minute and often unisexual; petals often 0; stamens 1–8; ovary inferior; fruit various. A small order of mostly aquatic plants, generally distributed.

1. HIPPURIS, L.

Flowers minute, solitary, axillary; petals 0; stamen 1; leaves linear, in whorls. Water-plants.

H. vulgaris, Mare's-tail; marshes and streams; not common.

2. MYRIOPHYLLUM, L.

Flowers minute, axillary, solitary or in spikes, unisexual; petals 2–4; stamens 2–8; styles 4; leaves very narrow. Water-plants.

Two species of Water-Milfoil—*M. verticillatum*, L., with the floral leaves all pinnatifid or pectinate, longer than the flowers; and *spicatum*, L., with the upper leaves entire, shorter than the flowers, occur in Switzerland; and the third British species, *M. alterniflorum*, DC., also in the Vosges and Pyrenees.

3. CALLITRICHE, L.

Flowers minute, solitary, axillary, unisexual; perianth 0; stamen 1; styles 2; leaves opposite, entire. Water-plants.

C. verna, L., Water-Starwort, with numerous sub-species; common in ponds and streams.

Order XXXIV.—LYTHRACEÆ.

Petals 3–6 or 0; calyx tubular, enclosing the ovary; stamens inserted on the calyx-tube; ovary 2–6-celled; leaves always entire, opposite or in whorls. A moderately large order, chiefly tropical; not alpine.

1. LYTHRUM, L.

Flowers axillary, usually bisexual; petals 4–6, red or pink; stamens 8–12; seed-vessel a 1–2-celled capsule.

L. Salicaria, L., Purple Loosestrife; flowers large, purple, in axillary whorls; stamens and styles of three different lengths; leaves often in whorls of 3; wet places, common. *L. Hyssopifolia*, L.; flowers solitary, small, pink, leaves usually alternate; wet places, rare.

2. PEPLIS, L.

Flowers minute, axillary; petals minute, purplish, or o.

P. Portula, L., Water-Purslane; stem prostrate, leaves obovate, obtuse; wet places or under water, local.

Order XXXV.—TAMARISCINEÆ.

Shrubs or small trees, with small flowers in dense solitary or panicled spikes; sepals and petals usually 5; stamens usually 10; ovary free, 1-celled; styles 2–5; seed-vessel a 2–5-celled capsule. A very small order; not alpine.

1. MYRICARIA, Desv.

Calyx 5-lobed; stamens 10, springing from beneath the ovary.

M. germanica, Desv.; flowers pink, in simple racemes, leaves small, linear, glaucous; sandy shores and beds of streams; Switzerland, Dauphiny, Pyrenees, not common.

Order XXXVI.—ONAGRACEÆ.

Flowers usually regular; calyx 2–4-lobed; petals 2–4; stamens 2–8; ovary usually 4-celled, with axile placentation, inferior; seed-vessel usually a 4-celled capsule. A large order, chiefly of the temperate regions.

1. EPILOBIUM, L.

Flowers solitary, axillary, or in terminal spikes, pink; calyx-limb 4-lobed; petals 4; stamens 8; ovary elongated, 4-celled, completely inferior; stigma 4-lobed; seeds tipped with a crest of long hairs. Many of the species of Willow-Herb are difficult to distinguish, and appear to hybridise freely.

A. Flowers irregular; petals undivided or emarginate; stamens and styles at length bent downwards:—*E. angustifolium*, L. (*spicatum*, Lam.), Great Willow-Herb, Rose-Bay; flowers numerous, very large (1 in. diameter), in a terminal spike, stem 2–4 ft., simple, leaves lanceolate, blue-green and veined on the under side; bushy places, common. *S. rosmarinifolium*, Hænke (*Dodonæi*, Koch); style hairy at the base, about as long as the stamens, leaves the same colour on both sides, and conspicuously veined; dry places; Switzerland, Jura, Dauphiny. *E. Fleischeri*, Hochst.; flowers large, few, calyx reddish-brown, stem 6–8 in., style pubescent, about half as long as stamens, leaves linear, the same colour on both sides; very high, local; Switzerland, Tirol, Dauphiny, Pyrenees.

B. Flowers regular; petals deeply bifid; stamens and styles erect; stigmas free, spreading:—*E. hirsutum*, L., Codlins-and-Cream; flowers rather large ($\frac{3}{8}$–$\frac{3}{4}$ in.), stem 3–4 ft., branched, very hairy, leaves opposite, half-amplexicaul, serrate, hairy; moist places, common. *E. parviflorum*, Schreb.; flowers numerous, small ($\frac{1}{4}$–$\frac{1}{3}$ in.), stem nearly simple, leaves not amplexicaul, alternate, whole plant usually pubescent; moist places, common. *E. montanum*, L., Common Willow-Herb; stem nearly simple, leaves opposite, stalked, ovate-lanceolate, usually glabrous; very common. *E. collinum*, Gmel.; resembling the last, but a smaller plant, with more branched stem and more numerous leaves; walls and rocks. *E. lanceo-*

latum, S. and M.; leaves oblong-lanceolate, on long stalks, mostly alternate, whole plant pubescent; Vosges, Puy-de-Dôme. *E. Duriæi*, Gay; stem prostrate and rooting at the base, leaves opposite, lanceolate, shortly stalked, with rounded base, dentate; Jura, Vosges, Puy-de-Dôme, Pyrenees.

C. Flowers regular; petals deeply bifid; stamens and styles erect; stigmas united into a club; no stolons:—*E. tetragonum*, L. (*adnatum*, Griseb.); stem with 2 or 4 raised lines, lanceolate, slightly decurrent, strongly denticolate, flowers small, purple; wet places. *E. Lamyi*, Schultz; closely resembling the last, but flowers rather larger, pink, plant bluish-green; wet places, local. *E. roseum*, Schreb.; flowers pink, stem with 2 or 4 raised lines, leaves thin, glabrous, narrowed at both ends, all distinctly stalked, buds inclined, acuminate; thickets. *E. alpestre*, Jacq. (*trigonum*, Schr.); leaves in whorls of 3 or 4, sessile, lanceolate, acuminate, flowers light pink, rather large, stem nearly simple; rocky; alpine.

D. Flowers regular; petals deeply bifid; styles and stamens erect, stigmas united into a club; stoloniferous: —*E. palustre*, L.; stem without raised lines, leaves narrowly lanceolate, usually opposite, nearly entire, sessile; wet places, common. *E. virgatum*, Fr. (*obscurum*, Rchb.); leaves lanceolate with rounded base, sessile, dentate, stem with 2 or 4 prominent lines; occasional. *E. alsinifolium*, Vill. (*origanifolium*, Lam.); flowers few, rather large, bright pink, leaves ovate, shining, with distant teeth, stolons underground, scaly; mountain streams, frequent. *E. anagallidifolium*, Lam.; leaves ellipticlanceolate, obtuse, entire, stalked, stolons above ground, leafy; moist, frequent. *E. alpinum*, L. (*nutans*, Schm.)

(Pl. 45); flowers small, few, leaves nearly sessile, lanceolate, obtuse, with a few teeth; damp; alpine, frequent.

2. CIRCÆA, L.

Flowers small, white; calyx bidentate, reflexed; petals
and stamens 2 each; stigma 2-lobed; leaves opposite.

The three British species of Enchanter's Nightshade—
C. Lutetiana, L., with ovate nearly entire leaves; *C.
alpina*, L., a smaller, more glabrous plant, the leaves
more cordate at the base and distinctly toothed; and *C.
intermedia*, Ehrh., intermediate between these—are met
with, the first generally distributed, the two latter in
more alpine situations; all sylvan plants.

3. ISNARDIA, L.

Flowers solitary, axillary; calyx-teeth, petals, and
stamens 4 each; seed-vessel an obovate 4-valved capsule;
seeds not crested. Not alpine.

I. palustris, L. (*Ludwigia palustris*, Ell.); a small glabrous marsh-plant, petals minute, red, or 0, leaves opposite, ovate, acute, entire, plant often tinged with red;
local.

4. TRAPA, L.

Flowers solitary, axillary; calyx-teeth, petals, and
stamens 4 each; calyx with 4 persistent lobes; fruit a
woody nearly globular spiny nut. Not alpine.

T. natans, L., Water-Nut; a submerged water-plant;
floating leaves rhomboidal, unequally dentate, submerged
leaves pinnatifid with capillary segments, flowers small,
white, fruit a woody nut with 2 or 4 prominent conical

spines; lakes in the southern district; rare. The remarkable fruit of the Water-Nut is used for making
rosaries, and is not unfrequently found among the debris
of the ancient lake-dwellings of Switzerland.

Order XXXVII.—CUCURBITACEÆ.

Prostrate or climbing herbs with simple or branched
tendrils; flowers unisexual; calyx 5 - lobed; petals 5,
usually very large; stamens 3, often with connate
anthers; ovary entirely inferior, 3-celled; fruit a 1-celled
many-seeded berry or "pepo," often very large and succulent. A rather large order, chiefly tropical and sub-
tropical; not alpine.

1. BRYONIA, L.

Flowers small, greenish-white; leaves 3–5-lobed or
angled; climbing by simple tendrils; berry small, globular.

B. dioica, L., White Bryony; diœcious, with red berries;
frequent in hedges.   *B. alba*, L.; monœcious, with black
berries; Western and Southern Switzerland; rare.

Order XXXVIII.—UMBELLIFERÆ.

Flowers always small, usually regular, sometimes unisexual, generally combined in simple or compound umbels,
with or without a whorl of bracts at the base of the compound umbel (general involucre) or of each of its divisions
(partial involucre); calyx-lobes 5 or calyx-limb 0; petals
5, sometimes unequal, usually white; stamens 5, epigy-
nous; fruit (cremocarp) consisting of two separable, dry

indehiscent carpels (mericarps) separated by a commissure, and each containing a single seed ; the pericarp often
deeply furrowed and furnished with vittæ or oil-glands ;
leaves almost always deeply and compoundly divided, and
with dilated leaf-stalk enclosing the stem.   A very large
order, belonging chiefly to the cooler parts of the Northern
and Western Hemisphere; but there are comparatively few
alpine species.   Owing to the many characters which are
common to the whole order, the distinctions between the
genera and species often depend on minute points.

Tribe HYDROCOTYLEÆ.—Umbels simple or flowers
very few ; fruit compressed laterally ; commissure narrow ;
vittæ o or obscure.

1. HYDROCOTYLE, L.

Flowers small, very few ; leaves entire or only lobed.
Not alpine.

H. vulgaris, L., Pennywort; with orbicular peltate
leaves and minute pinkish-green flowers; marshes and
bogs.

Tribe SANICULEÆ.—Umbels simple or very irregularly
compound ; fruit subterete or compressed dorsally ; commissure broad ; vittæ o or obscure.   Genera 2–5.

2. SANICULA, L.

Umbels small, crowded, nearly globular, with both
general and partial involucre ; calyx - teeth as long as
petals ; petals minute, deeply notched, with long inflexed

point; fruit ovoid, covered with hooked bristles. Not alpine.

S. europæa, L.; stem simple, nearly leafless, flowers often pink; woods, common.

3. HACQUETIA, DC.

Stem simple, leafless, bearing a single simple umbel with large involucre; flowers small, yellowish-green.

H. Epipactis, DC.; radical leaves 3–5-lobed, umbel capitate, surrounded by an involucre of yellowish bracts three times larger; bushy places; Carniola, Carinthia, Styria.

4. ASTRANTIA, L.

Umbels simple or compound; general involucre large, membranous, coloured; calyx - teeth longer than the petals; petals notched, with a long inflexed point; one vitta in each furrow.

A. major, L. (Pl. 46); radical leaves palmately divided, with elliptic ovate serrate segments, stem 1–2 ft., involucral bracts white, red, or green, about as long as umbel, calyx-teeth lanceolate; bushy places, common. *A. carinthiaca*, Hoppe; resembling the last, but stem more branched, involucral bracts nearly twice as long as umbel, usually bidentate at the tip; Carniola, Carinthia, local. *A. carniolica*, Wulf. (*gracilis*, Bartl.); involucral bracts shorter than umbel; calyx - teeth ovate, blunt; Carniola, Carinthia, Styria. *A. minor*, L.; radical leaves digitate, leaflets deeply cut, with narrow lanceolate segments, involucre and petals white, stem slender, 8–12 in.; alpine meadows, frequent. *A. alpina*, Schultz; radical leaves 3-partite to the base, unequally serrate,

involucral bracts white, as long as umbel, stem slender, 8–12 in.; bushy places; Tirol, Carniola.

5. ERYNGIUM, L.

Leaves fleshy, lobed or cut, with spiny teeth; flowers sessile, in very dense heads; involucral bracts hard, rigid; calyx-teeth longer than petals; petals narrow, with a long inflexed point.

E. alpinum, L.; radical leaves cordate, simple, stem-leaves 3-lobed, lobes deeply divided, involucral bracts digitate, with stiff teeth, blue or white; alpine pastures; Western and Southern Switzerland, Jura, Dauphiny, Carniola, Carinthia. *E. spina-alba*, Vill.; flowers white, leaves very thick and spiny, involucral bracts linear-lanceolate, pinnatifid, spiny, white; high; Dauphiny. *E. Bourgati*, Gou.; flowers blue, involucral bracts bluish-green, linear or linear-lanceolate, ending in a spine, and with 1–3 spiny teeth on each side; pastures; Pyrenees. *E. campestre*, L.; flowers white, involucral bracts nearly white, ending in a spine and toothed on each side, stem-leaves winged, decurrent, with auricled base; roadsides, occasional.

Tribe SMYRNIEÆ.—Umbels compound; fruit short, ovoid or didymous, compressed laterally, ridges not winged; vittæ obvious; commissure narrow; seeds grooved ventrally. Genera 6–9.

6. PHYSOSPERMUM, Cuss.

Flowers white; umbels compound; calyx-teeth small or o; petals with a long inflexed point; fruit didymous, bladdery; vittæ solitary in the furrows, leaves ternately compound.

P. aquilegiæfolium, Koch; flowers white, stem 1½–3 ft., glabrous, involucre small, bracts linear; alps of Dauphiny.

7. MOLOSPERMUM, Koch.

Leaves ter-pinnate; general involucre of few bracts or 0, partial of many bracts; petals with a long incurved point.

M. cicutarium, DC. (*Ligusticum peloponesiacum*, L.); fœtid, stem hollow, glabrous, 3–6 ft., leaflets ovate, acuminate, deeply pinnatifid, flowers white; Western and Southern Switzerland, Tirol, Carinthia, Dauphiny, Eastern Pyrenees.

8. PLEUROSPERMUM, Hoffm.

Leaves bi- or ter-pinnate; general and partial involucre of many bracts; petals without a recurved point; fruit inflated.

P. austriacum, Hoffm.; stem 2–4 ft., glabrous, striated, hollow, flowers rather large, white, leaves ter-pinnate, with elliptic dentate segments; pastures; Southern and Eastern Switzerland, Dauphiny.

9. CONIUM, L.

Leaves pinnately compound; general and partial involucre of many small bracts; calyx-teeth 0; petals obcordate; fruit crenulate, with many vittæ.   Not alpine.

C. maculatum, L., Hemlock; stem 3–6 ft., striated, hollow, with purple spots in the lower part; wood-sides and stony places; local; very poisonous.

Tribe AMMINEÆ.—Resembling the last, but the seeds flattened ventrally.   Genera 10–20.

10. SIUM, L.

Leaves pinnate, with toothed leaflets; bracts of general and partial involucre many; flowers white; calyx-teeth acute; petals with an inflexed point; vittæ many in each furrow.   Glabrous marsh plants; not alpine.

S. latifolium, L.; umbels terminal, lower leaves very large; and *angustifolium*, L. (*Berula angustifolia*, Koch); umbels opposite the leaves, leaves not so large; wet places, the latter very common.

11. HELOSCIADIUM, Koch.

Bracts of general involucre few or 0, of partial involucre many; flowers white; calyx-teeth 0; petals without an inflexed point; vittæ solitary in the furrows.   Glabrous marsh plants; not alpine.

H. repens, Koch; stem prostrate, umbels stalked, leaf-segments oval; and *nodiflorum*, Koch; umbels nearly sessile, leaf-segments oval-lanceolate; wet places and ditches, both rather local.

12. SISON, L.

Bracts of general and partial involucre few; flowers small, white; calyx-teeth 0; petals with an inflexed point; vittæ one in each furrow, very obscure.   Not alpine.

S. Amomum, L.; fœtid, stem 2–3 ft., very slender, solid, flowers very small and few, stem-leaves very small, 3-lobed or entire; road-sides, rare; Geneva, Carniola.

13. CICUTA, L.

Bracts of general involucre few or o, of partial involucre many, small; flowers white; calyx-teeth acute; petals with an inflexed point; vittæ one in each furrow, long. Not alpine.

C. virosa, L., Water-Hemlock; umbels opposite the leaves, root-stock hollow, chambered, leaves ter-pinnate, with lanceolate doubly serrate leaflets; ditches, occasional; very poisonous.

14. TRINIA, Hoffm.

Bracts of each involucre 1, 2, or o; flowers white; calyx-teeth small or o; petals entire or with a short incurved point; vittæ 1–2 in each furrow. Not alpine.

T. vulgaris, DC.; glabrous, bracts 1 or o, leaves bi- or ter-pinnate, with linear segments; open hill-sides.

15. PTYCHOTIS, Koch.

Involucre variable; calyx-teeth 5; petals obcordate, with a short incurved point; vittæ one in each furrow. Not alpine.

P. heterophylla, Koch; flowers white, radical leaves pinnate, with nearly round incised segments, stem-leaves multifid, with linear segments; dry places; Geneva, Tirol, Pyrenees.

16. FALCARIA, Riv.

Bracts of each involucre many, linear; calyx-teeth 5 in hermaphrodite, o in male flowers; petals with an

inflexed lobe; vittæ one in each furrow, elongated. Not alpine.

F. Rivini, Host. (*vulgaris*, Bernh.); flowers small, white, leaves rather thick, radical ones stalked, entire or ternate, stem-leaves pinnatifid, with linear lobes; fields; Switzerland (rare), Pyrenees.

17. BUPLEURUM, L.

Flowers yellow; bracts of partial involucre usually coloured and resembling a corolla; calyx-teeth o; petals hooded, with an inflexed point; vittæ o, or 1–2 in each furrow; leaves simple, entire. Mostly alpine.

A. General involucre o:—*B. rotundifolium*, L., Hare's Ear; partial involucre of 3–5 bracts, connate at the base, upper stem-leaves nearly round, perfoliate; fields, occasional.

B. General involucre present; bracts of partial involucre connate at the base:—*B. stellatum*, L. (Pl. 47); bracts of partial involucre 5–10, bright yellow, stem-leaves very few, linear-ovate, radical leaves linear-lanceolate; alpine pastures; Switzerland, Tirol, Carinthia, Dauphiny.

C. General involucre present; bracts of partial involucre free; leaves with one principal nerve:—*B. longifolium*, L.; bracts of partial involucre 5–7, ovate-acuminate, yellowish-green or brown, upper leaves cordate-amplexicaul; thickets; Switzerland (rare), Jura, Vosges, Dauphiny. *B. angulosum*, L. (*pyrenæum*, Gou.); bracts of partial involucre yellowish-green, nearly round, not mucronate, stem leafy, lower leaves linear-acuminate, upper cordate-amplexicaul, plant glaucous; Pyrenees.

D. General involucre present; bracts of partial involucre free; leaves with several nerves :—*B. falcatum*,
L.; flowers very small, bracts of partial involucre narrow,
awned, leaves all linear-lanceolate, often sickle-shaped;
hill-sides; Western Switzerland, Styria, Pyrenees. *B.
petræum*, L. (*graminifolium*, Vahl.); umbel solitary,
nodding, stem simple, leafless or with one leaf, radical
leaves linear, curved; very high; Southern Tirol, Carinthia, Dauphiny. *B. ranunculoides*, L.; bracts of partial
involucre broad, elliptic, umbel few-flowered, radical
leaves ovate or cordate, upper very narrow, stem 4–18
in., branched; dry alpine pastures. *B. caricifolium*,
Willd.; petals ovate, ochre-yellow, radical leaves linear,
fruit ribbed; Switzerland, local. *B. canalense*, Wulf.
(*caricifolium*, Rchb.); petals very small, pale yellow, fruit
not ribbed; Ticino, Carinthia, Pyrenees. *B. gramineum*,
Vill.; bracts of partial involucre linear-lanceolate, acuminate, umbel nodding, lower leaves linear-lanceolate, stem
leafy; Dauphiny, Pyrenees.

Several other species occur in the Pyrenean valleys.

18. CARUM, L.

Bracts of general and partial involucre few or 0; calyx-
teeth minute or 0; petals with a long inflexed tip; vittæ
1 in each furrow, elongated.   Not alpine.

C. Carui, L., Caraway; general and partial involucre
0 or of very few bracts, flowers white, outer larger and
irregular, leaves bi-pinnate, stem slender, branched, hollow;
waste places. *C. Bulbocastanum*, Koch (*Bunium Bulbocastanum*, L.), Pig-Nut; general and partial involucre
of 5–7 linear bracts, flowers white, outer rather larger,

leaves ter-pinnate, root a solitary tuber; Western and
Southern Switzerland, Carniola, Pyrenees.

19. ÆGOPODIUM, L.

Bracts of general and partial involucre few or 0; calyx-
teeth 0; petals broad, unequal, with an inflexed point;
leaves bi-ternate, leaflets broad; vittæ 0.   Not alpine.

Æ. Podagraria, L., Gout-Weed; leaves very large,
deltoid; waste places, common.

20. PIMPINELLA, L.

General involucre 0, partial involucre of a few bracts,
or 0; flowers white, pink, or yellow; calyx-teeth 0;
petals with a long inflexed point; vittæ numerous.

P. Saxifraga, L., Burnet Saxifrage; stem 1–3 ft.,
slender, terete, radical leaves pinnate, stem-leaves bi-
pinnate; *P. magna*, L.; stem 3–4 feet, angular, leaves all
pinnate; both in meadows and on hill-sides, common.
P. alpestris, Spreng., is a mountain variety of *Saxifraga*,
and *P. dissecta*, Retz., of *magna*; *P. rubra*, Hoppe, is a
form with pink flowers. *P. Tragium*, Vill.; stem shrubby
below, flowers white, umbel nodding, uppermost leaves
reduced to one, sheath surrounding the leaf-stalk; rocky
places; Dauphiny, Pyrenees.

Tribe SCANDICINEÆ.—Fruit elongated, compressed
laterally, commissure narrow; seed grooved ventrally.
Genera 21–25.

21. MYRRHIS, L.

Flowers white; bracts of general involucre few or 0,
of partial many, membranous; calyx-teeth minute or 0;

petals with a very short inflexed point; leaves decompound; vittæ 1 in each furrow or o; fruit almost winged.

M. odorata, L., Cicely; aromatic, leaves densely pubescent, fruit large, shining; mountain pastures, common.

22. CONOPODIUM, Koch.

Flowers white; general and partial involucre o or of
very few bracts; calyx-teeth o; petals of outer flowers
often irregular; vittæ several in each furrow. Not alpine.

C. denudatum, Koch (*Bunium flexuosum*, With.), Earth-
Nut; bracts o, leaves 3-ternate; dry banks; Dauphiny,
Pyrenees.

23. SCANDIX, L.

Flowers white; leaves bi- or ter-pinnate, with linear
segments; umbels simple; bracts of general involucre 1
or o, of partial involucre membranous; calyx-teeth small
or o; fruit very elongated. Not alpine.

S. Pecten Veneris, L., Shepherd's Needle, Venus's
Comb; a cornfield weed.

24. CHÆROPHYLLUM, L.

Leaves decompound; bracts of general involucre few
or o, of partial involucre many; fruit oblong or linear,
ribbed, not beaked; vittæ 1 in each furrow.

A. Petals not ciliate:—*C. temulum*, L.; leaves bi-pinnate, pubescent; hedge-rows, very common. *C. aureum*,
L.; leaves ter-pinnate, leaflets pinnatifid, stem thickened
at the joints, often spotted, glabrous; mountain woods.

B. Petals more or less ciliate:—*C. elegans*, Gaud.;
petals ciliate only at the tip, bracts of partial involucre
linear, membranous; rare; St. Bernard, Vorarlberg. *C.*

hirsutum, L. (*Villarsii*, Koch); petals completely ciliate, stem very hairy; Alps, Jura, Vosges, Dauphiny. *C. Cicutaria*, Vill.; flowers white or pink, petals ciliate all round, stem glabrous; damp meadows; Switzerland, local.

25. ANTHRISCUS, L.

Leaves decompound; bracts of general involucre few or 0, of partial involucre many; calyx-teeth minute or 0; fruit with a short beak, scarcely ribbed; vittæ solitary or 0.

A. Umbels with 8–15 rays:—*A. sylvestris*, Hoffm.; umbels on long stalks, leaves hairy, fruit glabrous; hedge-rows, very common. *A. alpina*, Jord.; leaves doubly pinnatifid, leaflets with linear segments; rocky, very rare; Bern. *A. nitida*, Gark.; leaves twice ternately pinnatifid, middle flowers of umbels barren; rare; Jura, Salève.

B. Umbels with 2–5 rays:—*A. vulgaris*, Pers.; leaves ter-pinnate, stem glabrous, flowers minute, fruit with hooked bristles; road-sides, Western and Southern Switzerland, rare. *A. Cerefolium*, Hoffm., Chervil; umbels sessile, stem hairy, fruit glabrous; rubbish-heaps. *A. fumarioides*, Spreng.; stem, leaves, and leaf-stalks ashen-grey, covered with silky hairs, ultimate segments of leaves very narrow; Carinthia, very rare.

Tribe SESELINEÆ.—Fruit globose or ovoid, not compressed laterally; commissure broad, lateral ridges usually distinct. Genera 26–44.

26. SESELI, L.

Bracts of general involucre various, of partial many, entire; flowers white; petals with a long inflexed point, notched; calyx-teeth small; fruit subterete, ridges not thickened or corky.

S. Libanotis, Koch (*Libanotis montana*, Crntz.), (including *daucifolium* and *gracile*); general involucre of many bracts, leaves bi-pinnate, bluish-green beneath, fruit rough with short hairs; mountain slopes. *S. montanum*, L.; general involucre o, bracts of partial involucre with narrow membranous edge, branches of umbel 6–10; calcareous hills. *S. annuum*, L. (*coloratum*, Ehrh.); branches of umbel 15–20, bracts of partial involucre with broad membranous edge; hill-sides, local. *S. tortuosum*, L.; resembling the last, but with tortuous stem, much branched from the base, branches of umbel 3–10; Tirol, Pyrenees. *S. caruifolium*, Vill.; stem slightly branched, solitary, bracts of partial involucre with broad membranous border, fruit glabrous; pastures; Dauphiny.

27. LIBANOTIS, All.

General and partial involucre of many bracts; calyx-teeth subulate; petals with a long inflexed point; vittæ 1, rarely 2, in each furrow.

L. athamantoides, DC.; lower leaves bi- or ter-pinnate, with ovate leaflets and mucronate segments, fruit glabrous; pastures; Carniola.

28. ATHAMANTA, Koch.

General involucre of few, partial of many bracts; petals hairy on the outside; fruit flask-shaped, hairy.

A. cretensis, L. (including *rupestris*, Scop.); leaves ter-pinnate, hairy or nearly glabrous, with linear segments, fruit ovate, covered with short distant hairs; pastures, frequent. *A. Matthioli*, Wulf.; leaves glabrous, fruit velvety; Tirol, Carniola.

29. TROCHISCANTHES, Koch.

Bracts of general involucre o or 1; leaves several times ternate; flowers greenish-white, the greater number barren; fruit slightly winged; vittæ 3-4 in each furrow.

T. nodiflorus, Koch; leaf-segments ovate-lanceolate, toothed, stem glabrous, striated, hollow, much branched; woods; Rhone Valley, Tirol, Dauphiny.

30. LIGUSTICUM, L.

General involucre various, partial of many bracts · flowers white or pink; calyx-teeth small or o; petals with a long inflexed point; fruit densely compressed; vittæ numerous; stem leafy.

L. ferulaceum, All., Lovage; stem 1-2 ft., bracts of general involucre numerous, pinnatifid; Jura, Salève, Dauphiny, rare. *L. Seguieri*, Koch; stem 2-4 ft., bracts of general involucre o or very few, entire; rare; Monte Generoso, Monte Baldo, Carniola.

31. PACHYPLEURUM, Led.

General involucre of many bifid or trifid bracts; flowers white or pink; petals with a long inflexed point; calyx-teeth o; leaves bi- or ter-pinnate; fruit much compressed; stem leafless.

P. simplex, Rchb. (*Gaya simplex*, Gaud.); leaf-segments linear, bracts of general involucre 7-10, flowers often dark pink, in crowded umbels; very high; Switzerland, Dauphiny, Pyrenees.

32. MALABAILA, Hacq.

General involucre of many bracts; calyx 5-toothed; leaves greatly divided, shining; fruit compressed; vittæ solitary.

M. Golaka, Hacq.; partial involucre of many bracts, stem branched, leaves ternately decompound, with doubly pinnatifid leaflets and pinnatifid segments; rare; Southern Tirol, Carniola.

33. MEUM, Jacq.

Bracts of general involucre few or o, linear, of partial 4–8; flowers white or pink; calyx-teeth o; petals with a short inflexed point; fruit compressed dorsally; vittæ numerous; fragrant.

M. athamanticum, Jacq.; Bald-Money (Pl. 48); leaf-segments capillary, almost verticillate, stem 6–12 in., flowers white; high pastures; Eastern Switzerland, Jura, Vosges, Dauphiny, Pyrenees. *M. Mutellina*, Gaert.; stem 12–18 in., leaf-segments linear-lanceolate, mucronate, flowers often reddish; high pastures. *M. pyrenaicum*, Gay (*Endressia pyrenaica*, Gay); umbels very small, nearly globular, lower leaves on long stalks; Eastern Pyrenees.

34. SELINUM, L.

Bracts of general involucre few or o, of partial many, small; flowers white or yellow; calyx-teeth o; fruit compressed dorsally; vittæ solitary. Not alpine.

S. caruifolium, L., Milk Parsley; stem angular, almost winged, leaf-segments mucronate, umbels corymbose; marshes, rare.

35. FŒNICULUM, Adans.

General and partial involucre o; calyx-teeth o; flowers yellow; petals with a short obtuse point, entire; fruit ovoid or oblong; vittæ solitary.   Not alpine.

F. officinale, All., Fennel; very glabrous, leaf-segments linear; stony places, occasional.

36. CNIDIUM, Cuss.

Involucre variable; calyx-teeth o; petals obovate, emarginate, with an inflexed point; fruit ovoid or oblong; vittæ solitary.

C. apioides, Spreng.; bracts of general and partial involucre, when present, linear, leaves ter-pinnate with ovate segments, leaf-sheaths loose, standing out from the stem; rare; Monte Generoso, Dauphiny.

37. DETHAWIA, Endl.

General involucre of 1–3, partial of many linear bracts; calyx-teeth 5, acute; petals elliptic, acute; fruit ovoid; vittæ solitary.

D. tenuifolia, Endl. (*Wallrothia tenuifolia*, DC.); leaves shining, ter-pinnate, with linear segments, stem flexuous, 4–15 in., glabrous, nearly leafless, slightly branched; rocky places; Central Pyrenees.

38. XATARDIA, Meissn.

General involucre o or of two bracts, partial of 4–12 linear bracts; calyx-teeth o; petals lanceolate, entire; vittæ solitary.

X. scabra, Meissn.; stem 6–8 in., very thick, glabrous,

hollow, nearly simple, flowers greenish-yellow, leaf-sheaths often very broad and tinged with violet; Eastern Pyrenees (rare), Vallée d'Eynes.

39. ŒNANTHE, L.

Bracts of general and partial involucre very few or 0; calyx-teeth acute; petals with a long inflexed point; stem hollow. Glabrous, mostly aquatic or marsh plants; not alpine.

Œ. fistulosa, L., Water-Dropwort; umbels with 2-4 branches, on hollow stalks, leaves on long stalks, with few narrow segments; ditches,rare. *Œ. Phellandrium*, Lam.; umbels with 7-10 branches, stem very thick, leaves often submerged, with capillary segments; ditches, rare. *Œ. Lachenalii*, Gmel.; leaves bi-pinnate, umbel with 8-15 branches, root-fibres tuberous; marshy meadows, rare. *Œ. peucedanifolia*, Poll.; resembling the last, but ultimate leaf-segments narrower, root-fibres not tuberous; ditches; Southern Switzerland, Jura, rare.

40. ÆTHUSA, L.

Bracts of general involucre 1 or 0, of partial about 3, always deflexed and on one side only; vittæ solitary. Not alpine.

Æ. Cynapium, L., Fool's Parsley; cultivated land, common.

41. SILAUS, Bess.

Bracts of general involucre few or 0, of partial many, small; flowers yellowish; calyx-teeth 0; petals with an incurved tip; fruit ovoid, oblong; vittæ obscure. Not alpine.

S. pratensis, L.; leaves ter-pinnate, with linear-lanceolate segments; damp meadows, common.

42. ANGELICA, L.

Bracts of general involucre few or O, of partial usually many; flowers white or pink; petals with a short inflexed point; leaves ternately bi-pinnate; fruit ovoid, compressed dorsally, with broad membranous wings; vittæ 1–2 in each furrow.

A. sylvestris, L.; leaf-segments oval or oval-lanceolate, toothed, with large sheaths; damp copses, common. *A. montana*, Schleich.; leaflets elliptic, sharply serrate, leaf-sheaths very large, inflated; bushy; alpine. *A. Razulii*, Gou.; flowers pink then white, stem 1½–3 ft., hollow, leafy, leaf-segments narrowly lanceolate, decurrent at the base; pastures; Pyrenees. *A. pyrenaica*, Spr.; stem 4–12 in., simple, furrowed, nearly leafless, leaf-segments linear, cuspidate; pastures; Pyrenees, Vosges.

43. ARCHANGELICA, Hoffm.

Bracts of partial involucre many; flowers greenish; petals entire; leaves ternately bi-pinnate; fruit broadly winged.

A. officinalis, Hoffm. (*Angelica Archangelica*, L.), Archangel; leaflets cordate-ovate, unequally serrate, leaf-sheaths connate, inflated; bushy; Styria, Carniola.

44. LEVISTICUM, Koch.

Bracts of general and partial involucre numerous; flowers yellow; calyx-teeth O; petals orbicular, with an inflexed point; fruit winged; vittæ solitary. Not alpine.

L. officinale, Koch; stem thick, hollow, 4–6 ft., leaves large, glabrous, bi- or ter-pinnate with rhomboidal segments; Tirol, Dauphiny, Pyrenees.

Tribe PEUCEDANEÆ.—Fruit much compressed dorsally, lateral ridges broadly winged; style short, stout. Genera 45–52.

45. PEUCEDANUM, L.

Involucre various; flowers white, yellow, or pink; calyx-teeth small; petals with an inflexed often bifid point; vittæ 1–3 in each furrow.

A. General involucre 0, or of not more than 3 bracts:
—*P. verticillare*, Koch (*Tommasinia verticillaris*, Bert.); stem 3–5 ft., much branched, upper branches in whorls, flowers greenish-yellow, fruit nearly globose; mountain slopes, local. *P. Ostruthium,* Koch (*Imperatoria Ostruthium*, L.); calyx-teeth 0, flowers white, leaves thick, leaflets ovate, dentate; bushy places, frequent. *P. angustifolium*, Rchb.; leaves thin, bi-ternate, leaflets connate, acuminate, incised, pinnatifid at the base; Ticino, rare. *P. caruifolium*, Vill. (*Chabræi*, Rchb.); leaves simply pinnate, segments pinnatifid with linear lobes, flowers greenish or yellowish; Jura, Pyrenees. *P. officinale*, L.; flowers yellow, stem 1½–3 ft., glabrous, leaf-segments linear-acuminate; marshes.

B. General involucre of 4 or more bracts; edges of carpels winged:—*P. austriacum*, Koch; stem furrowed, leaflets ter-pinnate, segments linear-lanceolate, flowers white; stony hills; Western and Southern Switzerland, Styria, Carinthia. *P. raiblense*, Koch; resembling the last, but leaf-segments narrowly linear; rare; Ticino, Southern Tirol, Carinthia, Carniola.

C. General involucre of four or more bracts; edges of carpels narrow; leaf-stalk triangular, channelled above :— *P. venetum*, Koch; flowers white, involucre spreading, umbels small, stem furrowed, leaf-segments linear-lanceolate; stony hills; Southern Switzerland, Tirol. *P. Orcoselinum*, Mönch.; flowers white, involucre reflexed, fruit orbicular, leaves green on both sides, segments deeply incised; dry hill-sides. *P. Cervaria*, Cuss.; flowers white or pink, involucre reflexed, fruit ovoid, leaves glaucous beneath, segments with mucronate teeth; dry hills. *P. alsaticum*, L.; flowers yellow, involucre spreading, fruit ovoid, leaves green on both sides; Jura, Pyrenees.

D. General involucre of four or more bracts, leaf-stalk cylindrical :—*P. palustre*, L.; flowers small, white, umbels large, fruit ovoid, on a long stalk, leaflets lanceolate, plant very glabrous; marshes.

46. PASTINACA, L.

General and partial involucre o; flowers small, white; calyx-teeth o; fruit with rather narrow wings; leaves simply pinnate. Not alpine.

P. sativa, L. (*Peucedanum sativum*, Benth.), Wild Parsnip; stem pubescent, leaflets in 2–5 pairs, broad, pubescent; waste places, common. *P. opaca*, Bernh.; resembling the last, but taller, the umbels smaller, pubescence denser; Switzerland.

47. HERACLEUM, L.

Involucre various; flowers white, pink, or yellowish; calyx-teeth small or o; petals generally unequal, with an inflexed point; fruit orbicular, with narrow wings; vittæ solitary, club-shaped, or o.

A. Outer flowers of umbel rayed; flowers white or pink; petals bifid; leaves pinnate, pinnatifid, or ternate :—
H. Sphondylium, L., Cow-Parsnip; leaves very large, coarse, pinnate, lobes of larger petals oval, spreading; meadows, very common (*H. elegans,* Jacq., is a mountain variety). *H. montanum,* Schleich. (*Panaces,* Rchb.); leaves glabrous or tomentose beneath, segments acuminate, lobes of larger petals linear, spreading; mountain pastures, rare; Zermatt, Ticino, Jura, Styria, Dauphiny, Pyrenees. *H. austriacum,* L.; leaves bi-pinnate, leaf-sheaths narrow, flowers white, fruit glabrous; Tirol, Styria, Carniola, Carinthia. *H. siifolium,* Rchb.; resembling the last, but flowers pink, fruit hairy; Carniola.

B. Leaves palmate; leaflets cordate at the base; flowers white; larger petals bifid:—*H. palmatum,* Baumg.; leaves 7–9-lobed, leaflets broad, acuminate, serrate, fruit warty; Tirol. *H. alpinum,* L.; leaves very large, 3–5 lobed, leaflets roundish; Jura, Valais. *H. pyrenaicum,* Lam.; lobes of larger petals oval, lower leaves orbicular in outline, leaves downy beneath; Pyrenees, Monte Baldo.

C. Flowers greenish-yellow; petals nearly alike :—
H. sibiricum, Koch (including *longifolium,* Jacq., and *flavescens,* Bess.); leaves pinnate, leaflets in about five pairs, serrate; alpine pastures.

D. Outer flowers of umbel rayed; vittæ o:—*H.*

minimum, Lam.; stem 4–12 in., procumbent, flowers white, leaves small, bi-pinnate; Dauphiny.

48. TORDYLIUM, L.

Bracts of general and partial involucre linear, small, or 0; petals with an incurved point, often unequal; fruit compressed dorsally; vittæ 1–3 in each furrow. No alpine.

T. maximum, L.; leaves simply pinnate, rough; flowers small, white or pink; plant hispid; Switzerland, rare (Canton de Vaud), Pyrenees.

49. DAUCUS, L.

Bracts, when present, pinnatifid; calyx-teeth narrow or 0; petals often unequal; umbel the shape of a bird's nest; fruit spiny; vittæ solitary, beneath the secondary ridges. Not alpine.

D. Carota, L., Wild Carrot; very common.

50. CAUCALIS, L.

Bracts of general involucre 0 or very few; flowers white or purplish; petals often unequal; leaves pinnate, hispid; ridges of fruit with 1–3 rows of spines; vittæ solitary beneath the ridges. Not alpine.

C. daucoides, L.; leaves bi- or ter-pinnate, spines of fruit hooked; fields, frequent. *C. latifolia*, L. (*Turgenia latifolia*, Hoffm.); very hispid, leaves simply pinnate, flowers pink, spines of fruit spreading, not hooked; fields, occasional.

51. TORILIS, Hoffm.

Resembling *Caucalis*, but fruit covered with bristles between the primary ridges.   Not alpine.

T. Anthriscus, Gmel.; bracts of involucre usually 7–9, spines of fruit not hooked; hedge-banks, very common. *T. infesta*, Hoffm. (*arvensis*, Huds., *helvetica*, Jacq.); bracts of general involucre 0 or 1, spines of fruit hooked; hedge-banks, occasional.

52. ORLAYA, Hoffm.

Resembling *Daucus*, but bracts entire; secondary ridges of fruit keeled, bearing 2–3 rows of subulate spines.

O. grandiflora, Hoffm.; outer petals very large, ten times as long as the inner ones, leaves bi or ter-pinnate, with linear segments; cornfields, especially at a high elevation.

Tribe THAPSIEÆ.—Fruit compressed dorsally; primary ridges filiform; some or all of the secondary ridges broadly winged.

53. LASERPITIUM, L.

Bracts of general and partial involucre usually many; calyx 5-toothed; petals obovate, emarginate, with an inflexed point; fruit with 4 or 5 broad wings; aromatic.

A. Leaves bi-pinnate only:—*L. prutenicum*, L.; flowers white, leaf-segments pinnatifid, with lanceolate lobes, stem deeply furrowed; damp meadows, local. *L. peucedanoides*, L.; flowers white, leaf-segments undivided, linear, fruit ovoid; Tirol, Carniola, rare. *L. Gaudini*, Mor.; general involucre 0, or of 1–3 leaves, petals yellow with a red border, stem round; rare; Switzerland (St. Gallen), Tirol.

B. Leaves ter-pinnate or decompound; flowers white:—
L. Siler, L.; leaves ter-pinnate, with lanceolate entire segments, glabrous, fruit linear; Switzerland, Jura, Dauphiny, Pyrenees. *L. latifolium,* L.; leaf-segments large, ovate, cordate at the base, glabrous or rough beneath, stem terete; stony hills. *L. hirsutum,* L. (*L. Panax,* Gou.); resembling the last, but leaflets small, ovate, with linear segments; dry, high; Switzerland, Carniola, Dauphiny. *L. Chironium,* Scop.; bracts of general involucre with white margin, stem angular, furrowed, leaf-sheaths inflated; Carniola, rare. *L. Nestleri,* Soy. (*aquilegiæfolium,* DC.); general involucre of 1–3 setaceous bracts, leaf-sheaths swollen; Pyrenees. *L. gallicum,* Bauh.; bracts of general involucre many, reflexed, linear-lanceolate, ciliate, bifid or trifid, leaves decompound, stem striated; Dauphiny, Pyrenees.

Order XXXIX.—ARALIACEÆ.

Flowers usually in umbels; calyx-teeth, petals, and stamens 5 each; ovary entirely inferior; fruit a drupe or berry, with one or more 1-seeded cells. A rather large order of erect or climbing shrubs or trees, chiefly tropical, with only one Central European representative.

1. HEDERA, L.

Climbing shrubs with simple leaves; flowers in umbels; fruit a black berry.

H. Helix, L., Ivy; rocks and walls.

Order XL.—CORNACEÆ.

Flowers small, regular; calyx-teeth 4-5, small, or 0; petals and stamens 4-5; ovary 1-4-celled, inferior; fruit a 1-4-stoned drupe. A very small order belonging to the North Temperate Zone.

1. CORNUS, L.

Flowers in dichotomous cymes or involucrate umbels; calyx-teeth, petals, and stamens 4 each; ovary 2-celled; fruit a drupe with a 2-celled stone.

C. sanguinea, L., Dogwood, Cornel; a shrub with opposite ovate leaves, white flowers in terminal cymes, and small black drupe; hedges and thickets. *C. mas*, L.; flowers yellow, appearing before the leaves, in simple umbels, surrounded by an involucre of four bracts, drupe red; mountain woods, frequent.

CLASS III.—COROLLIFLORÆ.

Corolla composed of petals more or less united at the base, rarely 0; stamens hypogynous or epipetalous. Orders XLI.—LXVIII.

Order XLI.—CAPRIFOLIACEÆ.

Flowers regular or irregular, collected into cymes; calyx-teeth 3-5; corolla sometimes 2-lipped; stamens 4-10; ovary entirely inferior; leaves opposite, simple or pinnate; fruit a berry or drupe. A small order, chiefly of the Northern Hemisphere, mostly shrubs.

1. SAMBUCUS, L.

Flowers small, in corymbose cymes; corolla regular, rotate or campanulate, 3–5-lobed; stamens 5; leaves pinnate; fruit a berry.

S. nigra, L., Elder; a small tree, stipules very small or o, corolla white, berry black; woods. *S. Ebulus*, L., Danewort; stem herbaceous, corolla white tipped with pink, stipules large, leafy, berry black; road-sides, local. *S. racemosa*, L.; stem woody, corolla greenish-yellow, flowers in ovoid panicles, stipules minute or o, berry red; mountain woods.

2. VIBURNUM, L.

Flowers small, in corymbose or globular cymes; corolla regular, rotate or tubular; calyx-teeth, corolla-lobes, and stamens 5 each; leaves simple; fruit a dry or fleshy drupe. Shrubs; not alpine.

V. Lantana, L., Wayfaring-Tree; leaves ovate-cordate, serrate, very pubescent beneath, without stipules, flowers small, white, drupe black; woods and hedges. *V. Opulus*, L., Guelder-Rose; leaves 3-lobed, nearly glabrous, stipulate, inflorescence nearly globular, outer flowers large, barren, with neither pistil nor stamens, drupe red; woods and thickets.

3. ADOXA, L.

Flowers small, green, 4–5 in a dense umbel; leaves ternate.

A. Moschatellina, Moschatel; a small glabrous herb, 2–4 in. high, with a musky odour, flowering in the very early spring, lower leaves bi-ternate, upper ternate;

woods and hedge-banks; Western Switzerland, Jura, Vosges, Dauphiny, Pyrenees, not common.

4. LINNÆA, L.

Flowers two on each stem, large, nodding, campanulate; stamens 4; stem trailing; leaves simple.

L. borealis, L. (Pl. 49); corolla white with red streaks, leaves thick, glabrous, evergreen, nearly orbicular, on short stalks. This very beautiful plant is found chiefly in fir-woods in Eastern and Southern Switzerland (Grisons, Valais), Tirol, Carinthia, and Salzburg.

5. LONICERA, L.

Shrubs with entire leaves, often climbing; flowers in pairs or in small cymes; calyx-teeth 5, often unequal; corolla tubular or campanulate, often 2-lipped; stamens 5; ovary 2–3-celled; fruit a 2–3-celled berry.

A. Stem climbing; flowers 2-lipped, with a very long tube, fragrant, in terminal heads or axillary cymes:—
L. periclymenum, Woodbine, Honeysuckle; cymes united into stalked terminal heads, flowers cream-coloured or red; woods; Switzerland (not common), Jura, Pyrenees. *L. Caprifolium*, L.; flower-heads sessile, flowers purple or yellowish-white, upper leaves connate at the base; Southern Switzerland (rare), Pyrenees. *L. etrusca*, Saut.; flower-heads on long stalks, flowers purple or orange, leaves not connate; Southern Switzerland, Tirol, Carniola, Pyrenees, rare.

B. Stem not climbing; flowers few, axillary, with a short tube, or campanulate:—*L. Xylosteum*, L., Fly-Honeysuckle; flowers two in each leaf-axil, yellowish-white,

XLIX.—LINNÆA BOREALIS.

ovaries connate at the base, tube of corolla short, gibbous, leaves pubescent, stalked, berries red; hedges.
L. nigra, L.; flowers pink, tube of corolla short, gibbous,
ovaries connate at the base, flower-stalk very long, leaves
nearly glabrous, berries black; mountain woods. *L.
alpigena*, L. (Pl. 50); flowers pale pink, tube of corolla
short, gibbous, flower-stalk very long, ovaries almost
completely connate, leaves acuminate, berries red; alpine
woods; in the Jura the red berries are very common in
the woods in autumn. *L. pyrenaica*, L.; flowers very
light pink, fragrant, tube campanulate, scarcely gibbous,
limb nearly regular, ovaries connate at the base, leaves
very shortly stalked, berries red; Pyrenees. *L. cærulea*,
L.; flowers yellowish, corolla campanulate, nearly regular, flower-stalk short, leaves blunt, ovaries nearly completely coherent, berries blue-black; thickets; Alps, Jura,
Pyrenees.

Order XLII.—RUBIACEÆ.

Ovary 2-celled, inferior; fruit 2-lobed, 2-seeded. All
the European genera belong to the tribe STELLATÆ:—
Flowers small, in axillary or terminal cymes; calyx 4–5-
toothed, or teeth 0; corolla 4–5-lobed; stamens 4; leaves
always forming a whorl, with or without the foliaceous
stipules. A very large tropical order, of which the
STELLATÆ are the only representatives in the Northern
Temperate Zone.

1. RUBIA, L.

Calyx-limb 0; flowers yellowish; corolla 5-lobed; fruit
a berry. Not alpine.

R. peregrina, L., Madder; leaves lanceolate, 4–6 in a

whorl, with reflexed bristles ; hedges ; Southern Switzerland (rare), Pyrenees.

2. GALIUM, L.

Flowers very small, white, red, or yellow, in axillary
or terminal cymes ; calyx-limb annular ; corolla usually
4-lobed; leaves (and stipules) in whorls of 4-12. Prostrate
or scrambling herbs with weak stem ; mostly lowland.

A. Flowers yellow :—*G. Cruciata,* Scop.; leaves 4 in a
whorl, elliptic, hairy ; and *verum,* L. ; Lady's Bedstraw,
leaves 8-12 in a whorl, linear; are common lowland
plants ; also *G. vernum,* Scop.; flowers few, leaves 4 in
a whorl, oval or lanceolate, ciliate ; stony places ; Ticino,
Pyrenees. *G. ochroleucum,* Wolf., is probably a hybrid.

B. Flowers red or purple :—*G. purpureum,* L. ; flowers
small, blood-red, in a branched panicle, flower-stalk deflexed, leaves 8-10 in a whorl, linear, mucronate; Southern
Switzerland, Carniola, Carinthia. *G. rubrum,* L.; flowers
large, wine-red, petals acuminate, leaves 8 in a whorl,
linear ; Switzerland, Tirol, Lombardy, rare.

C. Flowers white, usually few ; stem and leaves more
or less hispid ; not alpine :—*G. palustre,* L. (including
elongatum, Presl.); and *uliginosum,* L., are common
English marsh species. *G. Aparine,* L., Goose-Grass,
Cleavers, with very hispid stem, leaves, and fruit, is
very common in hedges. *G. tricorne,* L., with larger
fruit, and the fruit-stalks arched and reflexed, occurs
occasionally in hedges. *G. spurium,* L. (*Vaillantii,* DC.),
with very small greenish flowers and narrower leaves,
is found on cultivated land, especially among flax. *G.
parisiense,* L. (*anglicum,* Huds.); flowers greenish, leaves

mucronate, finally reflexed ; Western and Southern
Switzerland, Jura, Pyrenees. *G. saccharatum*, L.; flowers
often unisexual, fruit large, tubercular ; occasional.

D. Flowers white ; stem and leaves not hispid ; leaves
broader than linear :—*G. Mollugo*, L. (including *elatum*,
Thuill., and *erectum*, Huds.) ; and *saxatile*, L. ; are very
common English plants. *G. boreale*, L. ; leaves 4 in a
whorl, lanceolate, fruit hispid ; damp meadows and rocks.
G. rotundifolium, L. ; leaves 4 in a whorl, broadly ovate,
fruit rough ; mountain woods. *G. pedemontanum*, All. ;
leaves 4 in a whorl, elliptic, cymes axillary, stem prickly ;
sandy places ; Southern Switzerland, Piedmont. *G. tri-
florum*, Mich. ; leaves 6 in a whorl, oblong - lanceolate,
fruit small, prickly, plant fragrant ; woods ; Switzerland,
very rare. *G. sylvaticum*, L. ; leaves oblong-lanceolate ;
flower-stalk very slender, drooping, stem ridged ; woods.
G. aristatum, L. (*lævigatum*, L.); leaves narrower, corolla-
lobes acuminate - mucronate ; Pfeffers, Geneva, Ticino,
Dauphiny, Pyrenees. *G. helveticum*, Weig. ; leaves 7–8
in a whorl, oblong-lanceolate, mucronate, fleshy, flowers
almost concealed among the leaves, fruit large, fruit-
stalk deflexed ; Switzerland, Tirol, Styria, Dauphiny,
Pyrenees. *G. sylvestre*, Poll. (including *austriacum*,
Jacq. ; *commutatum*, Jord. ; *montanum*, Vill. ; *Lapeyrou-
sianum*, Jord.; *anisophyllon*, Vill.; and *tenue*, Vill.) ; leaves
variable, 7 – 8 in a whorl, usually oblinear - lanceolate,
mucronate, fruit-stalk not deflexed ; meadows. *G. myri-
anthum*, Jord. (including *obliquum*, Vill. ; *luteolum*, Jord. ;
and *leucophæum*, G. and G.) ; leaves variable, 9–12 in
a whorl, usually lanceolate - oblong, mucronate, softly
pubescent beneath, flowers more numerous ; Southern
Switzerland, Jura, Dauphiny. *G. megalospermum*, Vill.

(*Villarsii*, DC.); flowers in short terminal cymes, leaves 6 in a whorl, very fleshy, fruit large ; high ; Dauphiny, Pyrenees.

E. Flowers white ; stem and leaves not hispid'; leaves very narrow :—*G. rigidum*, Vill. (*cinereum*, Gaud.); resembling *Mollugo*, but the leaves linear-subulate ; Switzerland, very rare. *G. alpicola*, Jord. ; cyme narrow, pyramidal, leaves 8–9 in a whorl, linear, mucronate, reflexed, stem somewhat hispid towards the summit ; high ; Dauphiny. *S. intertextum*, Jord. ; cyme broad, spreading, leaves 7–9 in a whorl, linear, mucronate, reflexed, somewhat warty ; Dauphiny. *G. argenteum*, Vill. ; leaves 6–8 in a whorl, linear, covered with silvery glands, stem flexuose ; high ; Dauphiny. *G. pusillum*, L. (*pumilum*, Lam. ; *hypnoides*, Vill.); cyme short, leaves linear, with a long awn, sometimes hispid, plant cæspitose ; Pyrenees, Styria. *G. cæspitosum*, Ram. ; very cæspitose ; flowers in small cymes, almost hidden among the leaves, leaves 6–8 in a whorl, linear, awned; Pyrenees. *G. pyrenaicum*, Gou. ; cæspitose, flower-stalks shorter than the leaves, 1-flowered, leaves 6 in a whorl, linear, with long awns ; Pyrenees. *G. margaritaceum*, Kern. ; Southern Tirol ; and *baldense*, Spreng., Switzerland, rare ; probably belong to this section, but are insufficiently described.

F. Flowers white, with a distinct tube :—*G. glaucum*, L. ; leaves 6–8 in a whorl, linear, flowers in long-stalked cymes ; stony places ; Dauphiny, Pyrenees.

3. ASPERULA, L.

Flowers very small, white, blue, or pink, in axillary or terminal cymes ; calyx-limb 4-toothed ; corolla funnel- or bell-shaped, 4-lobed ; stamens 4. Not alpine.

A. Flowers white :—*A. odorata*, L., Woodruff; leaves
6–8 in a whorl, lanceolate, ciliate, fruit hispid, plant
fragrant; woods, common. *A. 'taurina*, L.; flowers
crowded, fragrant, subtended by ciliate bracts, leaves 4
in a whorl, lanceolate, ciliate, fruit glabrous; woods;
Southern Switzerland, Jura, Dauphiny. *A. glauca*, Bess.
(*galioides*, M.B.); leaves 8 in a whorl, linear, stiff,
glaucous; Switzerland, Jura, Styria, rare.

B. Flowers light pink, occasionally white :—*A. cynan-
chica*, L.; flowers few, small, funnel-shaped, stem prostrate, leaves 4 in a whorl, linear, fruit tubercular; hill-
sides, especially calcareous, common. *A. flaccida*, Ten.
(*longiflora*, W.K.); stem 1–2 ft., corolla-tube longer than in
the last, about twice as long as limb; Ticino. *A. montana*,
Willd.; corolla-tube three times as long as limb, leaves
shorter; Ticino. *A. tinctoria*, L.; corolla often 3-lobed,
stem usually solitary, erect; Western and Northern Switzerland, Jura, Vosges, Carniola, Pyrenees, rare. *A.
hirta*, Ram.; corolla-tube somewhat longer than limb,
leaves 6 in a whorl, linear, ciliate, plant hairy; Pyrenees.

C. Flowers blue :—*A. arvensis*, L.; corolla-tube about
as long as limb, bracts ciliate, leaves 6–8 in a whorl,
linear; cultivated land; Western Switzerland, Pyrenees.

4. SHERARDIA, L.

Flowers small, blue or pink, in terminal heads surrounded by bracts; corolla funnel-shaped; fruit didymous,
crowned by the enlarged calyx-limb. Not alpine.

S. arvensis, L.; flowers pink, leaves 4–6 in a whorl,
lanceolate, awned; cultivated land, common.

Order XLIII.—VALERIANACEÆ.

Flowers in compound cymes, often unisexual, usually irregular; calyx-limb often developing into a feathery pappus; calyx-lobes, corolla-lobes, and stamens 1–5 each; ovary inferior, 3-celled; leaves opposite.   A small order, belonging chiefly to the Northern Temperate Zone.

1. CENTRANTHUS, L.

Calyx-limb developing into a feathery pappus; corolla 5-lobed, irregular, spurred at the base; stamen 1.

C. angustifolius, DC.; flowers pink or white, spur of corolla about as long as ovary; leaves linear or linear-lanceolate; rocky places; Jura, Pyrenees.   *C. ruber*, DC., Red Valerian; spur slender, upper leaves triangular, ovate, plant very glabrous; Ticino, Lombardy.

2. VALERIANA, L.

Calyx-limb developing into a feathery pappus; corolla 5-lobed, nearly regular or gibbous at the base; stamens 3.

The two English species of Valerian, *V. officinalis*, L., with bisexual, and *dioica*, L., with unisexual flowers, are common in damp places.   The following also occur:—

V. sambucifolia, Mik.; flowers pale pink, all the leaves pinnate; wet places, frequent.   *V. supina*, L. (Pl. 51); flowers pink, in a dense terminal capitate cyme, stem 1–3 in., leafless except beneath the flowers, leaves spathulate, ciliate, entire or slightly toothed; wet rocks; Switzerland, rare (Münsterthal, Val Muschem), Tirol, Styria,

Carinthia, Salzburg. *V. saliunca*, All.; flowers light pink, fragrant, in a dense terminal capitate cyme, stem 2-6 in., leafy, leaves lanceolate, entire, glabrous; high; Western Switzerland, Jura, Tirol, Piedmont, Dauphiny, rare. *V. celtica*, L.; flowers yellowish, in small axillary cymes, forming an interrupted spike, stem 2-8 in., leaves linear-lanceolate, entire, whole plant fragrant; high rocks; local. *V. saxatilis*, L.; flowers white, in loose panicles, stem 4–12 in., leafless, radical leaves lanceolate, entire, on long stalks; high; Eastern Switzerland, Lombardy. *V. elongata*, L.; flowers yellow, in a loose panicle, radical leaves on long stalks, stem-leaves in about two pairs, ovate, sessile, coarsely toothed, stem about 1 ft., slender; high; Southern Tirol, Carniola, Carinthia, Styria. *V. tripteris*, L.; flowers white or pink, fragrant, crowded in a terminal cyme, stem 1-2 ft., leafy, leaves of the barren shoots cordate, of the flowering shoots pinnate with 3–5 leaflets; bushy; alpine. *V. montana*, L.; flowers pink or white, in a crowded corymbose cyme, stem 1-2 ft., leaves of the barren shoots rounded at the base, of the flowering shoots entire, rarely 3-cleft; high, moist. *V. pyrenaica*, L.; flowers pink, in corymbose cymes, stem 2-4 ft., hollow, radical leaves very large, cordate, coarsely and unequally toothed, stem-leaves of 3 leaflets; Pyrenees. *V. tuberosa*, L.; flowers pink, unisexual, lower leaves elliptical, stalked, upper pinnate, with 3–4 pairs of linear segments, stem 4–8 in., root tuberous; Carinthia, Pyrenees. *V. globulariæfolia*, Ram.; resembling the last, but root not tuberous, bracts linear-lanceolate, ciliate; Pyrenees.

3. VALERIANELLA, Tourn. (*Fedia*, Rchb.).

Flowers very small, solitary or in axillary cymes; corolla regular, funnel-shaped, 5-lobed; stamens 3. Small annual weeds in cultivated land; not alpine.

The English species of Lamb's-Lettuce or Corn-Salad, *V. olitoria*, Mœnch., *carinata*, Lois., *Auricula*, DC., and *dentata*, Poll. (*Morisonii*, DC.), occur in similar situations in Central Europe, and are difficult to distinguish.

Order XLIV.—DIPSACACEÆ.

Flowers small, collected into a dense spike or capitule, subtended by an involucre of bracts; calyx-limb cup-shaped, entire or 5-lobed, surrounded by an involucel; corolla funnel-shaped, 4–5-lobed; stamens 4, free; ovary 1-celled with one pendulous ovule; fruit indehiscent; leaves opposite or whorled. A small order, chiefly Asiatic.

1. CEPHALARIA, Schrad.

Capitule hemispherical; involucre composed of many soft green undivided bracts; involucel cup-shaped, with eight bristly teeth; receptacle covered with soft scales.

C. alpina, Schrad.; flowers pale yellow, stem 3–4 ft., leaves pinnate, leaflets 9–15; Southern Switzerland, Jura, Dauphiny, Pyrenees. *C. pilosa*, G. & G.; flowers white, involucel many-toothed, leaves pinnatifid at the base only, stem spiny in the upper part; Jura, Pyrenees. *C. leucantha*, Schrad.; flowers white, radical leaves simple, stem-leaves pinnatifid, capitule globular, plant almost shrubby; Pyrenees.

2. DIPSACUS, Tourn.

Capitule oblong or cylindrical; involucre composed of many very rigid, often spiny, bracts; leaves often connate at the base. Not alpine.

D. sylvestris, L., Wild Teasel; flowers purplish, leaves simple, obovate-lanceolate, sessile, prickly; road-sides. *D. pilosus*, L.; flowers nearly white, leaves stalked, with two small segments at the base; damp hedges. *D. laciniatus*, L.; flowers nearly white, stem-leaves pinnatifid, bristly; Western Switzerland (rare), Jura, Dauphiny, Pyrenees.

3. SCABIOSA, L.

Capitule hemispherical or flat; involucral bracts green, soft, undivided; outer flowers often larger and irregular, inner flowers tubular; calyx-limb awned; corolla curved.

S. lucida, Vill.; flowers violet or pink, outer ones rayed, leaves shining, lower simple, lanceolate, crenate, upper deeply pinnatifid, leaflets serrate; alpine rocks and pastures, frequent. *S. graminifolia*, L.; flowers light violet, outer ones rayed, leaves linear, entire, silky; stony pastures; Ticino (rare), Carinthia, Dauphiny. *S. suaveolens*, Desf.; flowers light violet, outer ones rayed, leaves of barren shoots entire, of flowering shoots pinnate with linear entire segments, awns of calyx very long; Bâle, Schaffhausen, Carniola, Vosges. *S. Columbaria*, L.; flowers lilac, outer ones strongly rayed, lower leaves inciso-lyrate, upper pinnate with pinnatifid lobes, all finely pubescent, stalk of capitule very slender; dry hills, common. *S. ochroleuca*, L.; resembling the last, but

with yellow flowers; Switzerland. *S. affinis*, G. and G.;
resembling *Columbaria*, but capitule nearly globular;
Dauphiny. *S. gramontia*, L. (*agrestis*, W.K.; *pyrenaica*,
All.); leaves more finely divided, awns of calyx shorter;
Southern Switzerland, Tirol, Styria, Pyrenees. *S. succisa*,
L. (*Succisa arvensis*, Mœnch.), Devil's Bit; flowers blue,
outer ones not rayed, capitule nearly globular, leaves
simple, root premorse; marshes, common.

4. KNAUTIA, Coult.

Resembling *Scabiosa*, but awns of calyx deciduous.

K. arvensis, Koch, Field-Scabious; flowers violet-blue,
upper leaves pinnatifid; very common. *K. sylvatica*,
Duby; flowers reddish-blue, leaves nearly entire; woods.
K. transalpina, Christ; flowers pink or purple, outer
flowers not much larger, stem about 6 in., white with
rough hairs; Switzerland, rare. *K. longifolia*, Koch;
flowers lilac, outer ones much larger, stem 1–2 ft., glandular-viscid, leaves oblong-lanceolate; Jura, Tirol, Styria,
Pyrenees. *K. mollis*, Jord.; flowers purple, leaves pubescent, almost silky, flower-stalk covered with glandular
hairs; Dauphiny, Pyrenees.

Order XLV.—CAMPANULACEÆ.

Calyx 5-cleft; corolla regular, 5-cleft; stamens 5;
anthers connivent round the style; ovary inferior, 2–3-
celled; style 1; stigmas 2–8; fruit a two- or more-celled
berry or capsule. A large order, chiefly of Temperate
climates.

LIII.—CAMPANULA LONGIFOLIA.

1. CAMPANULA, L.

Flowers in spikes or racemes, not surrounded by an involucre, usually blue; calyx sometimes with five reflexed basal lobes; corolla campanulate, often nearly entire; filaments with broad dilated base; fruit a 3–5-celled capsule, dehiscing below the calyx-limb by pores or valves.

A. Corolla bearded within, light blue :—*C. barbata*, L.; flowers drooping, calyx with a reflexed appendage between the lobes, stem 6–8 in., 1–10 flowered, leaves lanceolate, entire, plant woolly; alpine pastures, frequent; Switzerland, Jura, Carpathians, Erzgebirge, Dauphiny. *C. Zoysii*, Wulf.; flowers erect, corolla inflated below, turbinate, with five triangular teeth, stem 2–4 in., nearly prostrate, 1–5 flowered, leaves broadly elliptical, entire; clefts of rocks, rare; Eastern Alps, Styria, Carinthia, Carniola.

B. Corolla not bearded, bright blue; calyx with five appendages :—*C. alpina*, L. (Pl. 52); flowers large, azure-blue, stem 1–6 in., lower leaves linear-spathulate, slightly crenate, upper linear-lanceolate, entire, plant woolly; rare; Eastern Alps, Tirol, Styria, Carinthia. *C. longifolia*, Lap. (*speciosa*, Pourr.) (Pl. 53); flowers large, calyx-teeth linear-lanceolate, ciliate, with short appendages, stem 4–12 in., leaves linear-lanceolate, entire, ciliate; Pyrenees. *C. Allionii*, Vill.; flowers large, solitary, erect, violet, corolla-lobes spreading, calyx-teeth lanceolate, leaves linear-lanceolate, not ciliate; Dauphiny, Piedmont.

C. Calyx not appendiculate; corolla not bearded; flowers large, sessile, in dense spikes, blue or yellow :— *C. spicata*, L.; flowers violet, in an interrupted spike, narrowed above, leaves oblong-lanceolate, plant hispid;

Valais, Ticino, Tirol, Dauphiny, Carinthia, Carniola. *C. thyrsoidea*, L. ; flowers dull yellow, in a dense conti- nuous spike, corolla hairy without, leaves lanceolate, entire, hairy like the stem, which is glandular-viscid below ; alpine pastures ; Switzerland, Jura, Carpathians. *C. petræa*, L. ; flowers light yellow, stem 4–12 in., leaves broadly lanceolate, serrate, tomentose beneath, bracts elongated ; rare ; Tirol, Dauphiny, Lombardy. *C. Cer- vicaria*, L. ; flowers blue, calyx-lobes short, obtuse, leaves narrowed into a leaf-stalk, whole plant hispid ; thickets ; Switzerland, Jura, Lombardy, Vosges, rare. *C. glome- rata*, L. ; flowers small, blue or white, in dense clusters, stem 4–12 in., lower leaves cordate or rounded, calyx- teeth lanceolate ; open hill-sides.

D. Calyx and corolla as the last ; flowers blue, stalked, in racemes or panicles, or solitary ; capsule inclined or pendant. *C. latifolia*, L., Giant Bell-Flower ; flowers very large, erect, solitary in the axils of the leaves, corolla-teeth lanceolate, stem 3–4 ft., leaves ovate-lanceo- late, acuminate, serrate ; mountain woods. *C. Trachelium*, L., Bell-Flower ; flowers 2 or 3 in the axils of the leaves, erect, smaller than the last, stem 1–3 ft., stem-leaves nearly triangular, coarsely dentate, plant hispid ; woods, frequent. *C. rapunculoides*, L.; flowers pendant, in long slender unilateral racemes, funnel-shaped, corolla-lobes recurved, calyx-lobes reflexed after flowering, stem-leaves ovate-lanceolate, toothed ; road-sides ; Switzerland, Dau- phiny. *C. bononiensis*, L.; flowers smaller (¾ in.), nearly sessile, leaves lanceolate, finely toothed, tomentose be- neath, stem 12–18 in., tomentose ; thickets ; Southern Switzerland, Dauphiny, rare. *C. rhomboidalis*, L.; stem simple, 12–18 in., radical leaves usually 0, stem-leaves

LV.—CAMPANULA PULLA.

sessile, ovate, with sharp teeth, calyx-teeth linear, two-thirds as long as corolla; Switzerland, Jura, Dauphiny, Pyrenees, rare. *C. lanceolata*, Lap.; resembling the last, but calyx-teeth only one-third as long as corolla, leaves crowded, finely ciliate; pastures; Pyrenees. *C. rotundifolia*, L., Harebell; root-leaves cordate or reniform, crenate, stem-leaves linear-lanceolate; common. *C. carnica*, Sch. (Carniola, Carinthia), is probably only a mountain form of *rotundifolia*. *C. linifolia*, Lam. (Pl. 54); resembling *rotundifolia*, but calyx-teeth hardly half as long as corolla, stem-leaves linear, stem and leaves glandular-hairy; Dauphiny, Pyrenees. *C. Scheuchzeri*, Vill.; similar, but a larger plant with larger flowers, radical leaves elliptic or lanceolate; pastures, frequent. *C. pusilla*, Hænk.; cæspitose, flowers light blue, nodding, nearly hemispherical, stem 1- or very few-flowered, 2–4 in., lower leaves spathulate, upper linear-elliptical, plant nearly glabrous; rocks, frequent. *C. cæspitosa*, Scop.; very similar, corolla longer, dark blue, plant more hairy; local. *C. tenella*, Jord. (Dauphiny), and *Mathoneti*, Jord. (Dauphiny), are scarcely distinguishable from the preceding. *C. pulla*, L. (Pl. 55); flowers dark blue, solitary, nodding, stem 2–6 in., leaves all lanceolate or ovate, crenate, lower ones stalked; high; Styria, Carinthia, Carniola. *C. excisa*, Schleich.; corolla-lobes incised at the base and separated by a rounded sinus, calyx-teeth reflexed; very high; Southern Switzerland, rare (Simplon, Furka). *C. carpathica*, Jacq. (Pl. 56); flowers large, blue or white, funnel-shaped, on long stalks, stigmas sometimes 4, stem 6–12 in., radical leaves lanceolate with rounded leaves, crinkled; Carpathians.

E. As the last, but capsule erect :—*C. Rapunculus*, L.,

Rampions; flowers in an erect pyramidal panicle, blue,
calyx-teeth very long, subulate, corolla deeply divided,
leaves mostly ovate, on long stalks; road-sides, frequent.
C. patula, L.; flowers violet, funnel-shaped, with spreading lobes, on long slender stalks, calyx-teeth linear, leaves
mostly lanceolate, glabrous; hedge-rows; frequent. *C.
persicifolia*, L.; flowers few, large, in a loose raceme,
calyx-teeth lanceolate, corolla not deeply divided, with
rounded lobes; woods and meadows, not uncommon.
C. cenisia, L.; flowers small, about $\frac{1}{3}$ in., corolla funnel-
shaped, divided half-way down, dark blue, solitary, calyx
hairy, stem 2–3 in., leaves small, obovate, entire, ciliate;
very high; Switzerland, Tirol, Dauphiny. *C. Morettiana*,
Rchb.; flowers violet, solitary, $\frac{3}{4}$–1 in., campanulate,
corolla divided half-way down, leaves small, ovate, stalked,
serrate; clefts of rocks; Tirol, rare. *C. Raineri*, Perp.;
flowers large, blue, $\frac{3}{4}$–1 in., solitary or very few, campanulate, corolla divided half-way down, leaves spathulate,
crenate, stalked; rocks; Southern Switzerland, Tirol,
rare (Monte Generoso).

2. ADENOPHORA, Bess.

Resembling *Campanula*, but with a tubular hypogynous disk at the base of the style.

A. liliifolia, Bess.; styles much exceeding corolla; rare;
Ticino, Tirol, Lombardy, Carniola.

3. SPECULARIA, Heist.

Resembling *Campanula*, but ovary very long and slender; corolla rotate. Not alpine.

S. Speculum, DC.; flower large ($\frac{1}{2}$–$\frac{3}{4}$ in.), violet-purple,

LVI.—CAMPANULA CARPATHICA.

calyx-lobes linear; cultivated land; Switzerland. *S. hybrida*, DC.; flowers smaller ($\frac{1}{4}$ in.), lilac, calyx-lobes lanceolate, longer than corolla; cultivated land; Bâle, Schaffhausen.

4. JASIONE, L.

Flowers small, blue, in terminal heads surrounded by an involucre; corolla regular, 5-cleft to the base; anthers partially connate. Not alpine.

J. montana, L.; flowers lilac-blue, bracts of involucre entire, leaves obovate-oblong, plant pubescent; wood-sides, common. *J. perennis*, Lam.; flowers blue, bracts of involucre serrate, leaves oblong-lanceolate; Pyrenees, Jura, Vosges.

5. EDRAIANTHUS, DC.

Flowers blue, in terminal heads, surrounded by an involucre; corolla campanulate, 5-lobed.

E. croaticus, Kern.; stems 3–4 in., cæspitose, corolla $\frac{3}{4}$ in. long, leaves linear; pastures; Carniola, rare (Schnee-berg).

6. PHYTEUMA, L.

Flowers usually blue, in dense heads or spikes, sur-rounded by an involucre of bracts; corolla curved in bud, 5-cleft, with linear segments, anthers free; stigmas 2–3. An almost exclusively alpine genus, especially charac-teristic of Switzerland and Tirol.

A. Flowers stalked :—*P. comosum*, L.; corolla inflated below, large, lilac with dark violet tip, stem 2–6 in., leaves lanceolate, coarsely toothed; clefts of rocks; South-ern Tirol, Carinthia, Carniola, Lombardy, Carpathians.

B. Flowers sessile, in nearly hemispherical heads:—
P. Scheuchzeri, All. (*Charmelii*, Vill.) ; flowers lilac, bracts
of involucre linear, much longer than the inflorescence,
stigmas 2, stem slender, 12–18 in., radical leaves linear,
crenate or serrate, on long stalks, sometimes cordate,
stem-leaves linear ; alpine, rocky. *C. pauciflorum*, L. (Pl
57) ; flowers blue, 5–7 in each capitule, stem 2–3 in., involucral bracts oval, blunt, leaves obovate-lanceolate, slightly
crenate ; alpine pastures, frequent. *P. humile*, Schleich.
(Pl. 58) ; flowers violet, ten or more in an inflorescence,
bracts nearly as long as flowers, ovate-lanceolate, usually
dentate, stem 2–4 in., leaves linear, ciliate, calyx often
coloured ; very high ; Southern Switzerland, Tirol, Piedmont, Carinthia, Salzburg, rare. *P. hemisphæricum*, L. ;
flowers blue, elongated, inflorescence about 10-flowered,
bracts acuminate, about half as long as flowers, stem
2–4 in., leaves linear, glabrous ; very high, local. *P. latifolium*, Heuff. (*confusum*, Kern.) ; resembling the last, but
bracts acute or obtuse, not acuminate, leaves spathulate,
slightly crenate ; very high ; Carinthia, Styria. *P. Sieberi*,
Spreng. ; flowers violet, inflorescence rather longer, bracts
ovate, apiculate, serrate, stem 2–6 in., leaves ovate-lanceolate, crenate ; very high ; Tirol, Styria, Carniola, Carinthia.
P. orbiculare, L. ; flowers blue, stigmas 3, exserted,
bracts lanceolate, serrate, stem 6–12 in., often hollow,
leaves very variable in width, crenate-serrate ; pastures,
common.

C. Flowers sessile, numerous, in elliptical or cylindrical spikes:—*P. spicatum*, L. ; flowers yellowish-white or
rarely blue, stigmas 2, exserted, stem 1–2 ft., leaves
cordate-ovate, often spotted with brown ; woods and
thickets, frequent. *P. Halleri*, All. ; flowers dark blue

LVIII.—PHYTEUMA HUMILE .

or violet, filaments woolly, stem 2–3 ft., hollow, lower leaves cordate-ovate, upper lanceolate, plant glabrous; alpine pastures, frequent. *P. scorzonæræfolium*, Vill. (*Michelii*, All.); flowers blue, stigmas generally 2, stem 1–2 ft., glabrous, leaves glabrous or ciliate, linear-lanceolate; pastures; Southern Switzerland (Splügen), Tirol, Dauphiny, local. *P. betonicæfolium*, Vill.; flowers blue, stigmas generally 3, stem 8–18 in., leaves usually hairy, lower lanceolate or cordate-lanceolate, on long stalks; pastures; Switzerland, Piedmont, Dauphiny, Pyrenees, frequent. *P. nigrum*, Schmidt; flowers dark blue, otherwise like *spicatum*; Jura (Val de Joux), Pyrenees, rare.

Order XLVI.—AMBROSIACEÆ.

Flowers unisexual; male flowers in globular capitules, corolla regular, 5-lobed; female flowers solitary or in pairs, enclosed in an involucre, corolla 0. A small order, not alpine.

1. XANTHIUM, Tourn.

Female flowers in pairs; seed-vessel a spiny achene.

X. strumarium, L.; flowers green, seed-vessel covered with hooked prickles, leaves cordate, stalked, not spiny; rubbish heaps.

END OF VOL. I.

Printed by BALLANTYNE, HANSON & Co.
Edinburgh and London

www.ingramcontent.com/pod-product-compliance
Lightning Source LLC
Chambersburg PA
CBHW021507210326
41599CB00012B/1166